£5·50

PROGRESS IN MATERIALS SCIENCE

Volume 17

Edited by

BRUCE CHALMERS
Division of Engineering and Applied Physics,
Harvard University, Cambridge, Mass., U.S.A.

J. W. CHRISTIAN
Department of Metallurgy,
University of Oxford

and

T. B. MASSALSKI
Mellon Institute of Science,
Carnegie-Mellon University, Pittsburgh, Pa., U.S.A.

THE STORED ENERGY
OF COLD WORK

M. B. BEVER
*Department of Metallurgy and Materials Science,
Massachusetts Institute of Technology, Cambridge, Mass., U.S.A.*

D. L. HOLT
*Department of Mechanical Engineering,
University of Auckland, Auckland, New Zealand*

and

A. L. TITCHENER
*Department of Chemical and Materials Engineering,
University of Auckland, Auckland, New Zealand*

PERGAMON PRESS

OXFORD · NEW YORK · TORONTO
SYDNEY · BRAUNSCHWEIG

Pergamon Press Ltd., Headington Hill Hall, Oxford

Pergamon Press Inc., Maxwell House, Fairview Park, Elmsford, New York 10523

Pergamon of Canada Ltd., 207 Queen's Quay West, Toronto 1

Pergamon Press (Aust.) Pty. Ltd., 19a Boundary Street, Rushcutters Bay, N.S.W. 2011, Australia

Vieweg & Sohn GmbH, Burgplatz 1, Braunschweig

Copyright © 1973 Pergamon Press Ltd.

All Rights Reserved. No part of this publication may be reproduced, stored in a retrieval system, or transmitted, in any form or by any means, electronic, mechanical, photocopying, recording or otherwise, without the prior permission of Pergamon Press Ltd.

First edition 1973

Library of Congress Catalog Card No. 72-85096

Printed in Great Britain by Page Bros (Norwich) Ltd., Norwich

ISBN 0 08 017011 0

THE STORED ENERGY OF COLD WORK

M. B. BEVER
Department of Metallurgy and Materials Science,
MIT, Cambridge, Mass.

D. L. HOLT
Department of Mechanical Engineering,
University of Auckland, N.Z.

and

A. L. TITCHENER
Department of Chemical and Materials Engineering,
University of Auckland, N.Z.

CONTENTS

	Page
INTRODUCTION	
The nature of the stored energy of cold work	5
1. THERMODYNAMIC ASPECTS OF DEFORMATION PROCESSES	6
1.1. The work of deformation	7
1.2. Thermal effects of deformation	9
1.2.1. Elastic deformation	9
1.2.2. Plastic deformation	9
1.3. Changes in thermodynamic functions with deformation	10
1.3.1. Internal energy	10
1.3.2. Free energy	10
1.3.3. Entropy	11
1.3.4. Heat capacity	12
1.4. The energies and entropies of imperfections	12
1.4.1. Dislocations	12
1.4.2. Point defects	13
1.5. Irreversible thermodynamics and plastic deformation	14
2. METHODS OF MEASURING THE STORED ENERGY	16
2.1. Single-step methods	16
2.2. Two-step methods	19
2.2.1. Anisothermal annealing methods	19
2.2.2. Isothermal annealing methods	21
2.2.3. Reaction methods	22

3. THE EFFECT OF VARIABLES ON THE AMOUNT OF THE STORED ENERGY	23
3.1. Variables related to the deformation process	23
3.1.1. The extent of deformation	23
3.1.1.1. The ratio of the stored to the expended energy	49
3.1.2. Temperature of deformation	51
3.1.3. Rate of deformation	56
3.1.4. The deformation process	59
3.1.5. Deformation history	61
3.1.5.1. The Bauschinger effect	62
3.1.5.2. Work softening	62
3.1.6. Cyclic deformation (fatigue)	63
3.2. Variables related to the metal	66
3.2.1. Grain structure	66
3.2.1.1. Single crystals	69
3.2.2. Purity	73
3.2.3. The identity of the metal	76
3.2.4. Alloys and intermetallic compounds	80
3.2.4.1. Alloys with long-range order	85
3.2.4.2. Two-phase alloys	88
4. THE EFFECT OF VARIABLES ON THE RELEASE OF THE STORED ENERGY	89
4.1. Variables related to the deformation process	91
4.1.1. Extent of deformation	91
4.1.2. Temperature of deformation	93
4.1.3. Rate of deformation	96
4.1.4. The deformation process	96
4.1.5. Deformation history	97
4.1.6. Cyclic deformation (fatigue)	98
4.2. Variables related to the metal	99
4.2.1. Grain structure	99
4.2.1.1. Single crystals	99
4.2.2. Purity	100
4.2.3. The identity of the metal	103
4.2.3.1. Copper and nickel	103
4.2.3.2. Silver and gold	103
4.2.3.3. Aluminum	103
4.2.3.4. Other metals	105
4.2.4. Alloys	106
4.3. Variables related to the annealing process	110
4.3.1. Heating rate in anisothermal annealing	110
4.3.2. Temperature in isothermal and isochronal annealing	110
4.4. Energy release immediately following deformation	110
5. OTHER PROPERTIES MEASURED IN INVESTIGATIONS OF THE STORED ENERGY	112
5.1. Hardness	113
5.2. Flow stress	115
5.3. Resistivity	115

5.4. Density	117
5.5. The relation of the stored energy to the flow stress, resistivity and density	118

6. INTERPRETATION ... 122
  6.1. Historical survey ... 123
    6.1.1. Interpretation based on a phase change ... 123
    6.1.2. Interpretation based on functional relations ... 124
    6.1.3. Interpretation based on lattice strains ... 125
    6.1.4. Interpretation in terms of lattice imperfections ... 126
  6.2. The relation between variables and the structure of cold-worked metals ... 127
    6.2.1. Extent of deformation ... 127
      6.2.1.1. Recovery ... 128
    6.2.2. Temperature of deformation ... 128
    6.2.3. Rate of deformation ... 128
    6.2.4. The deformation process ... 128
    6.2.5. Deformation history ... 129
      6.2.5.1. Bauschinger effect ... 129
      6.2.5.2. Work softening ... 129
    6.2.6. Cyclic deformation (fatigue) ... 129
    6.2.7. Grain structure ... 129
      6.2.7.1. Single crystals ... 129
    6.2.8. Purity ... 130
    6.2.9. The identity of the metal ... 130
    6.2.10. Alloys and intermetallic compounds ... 130
      6.2.10.1. Alloys with long-range order ... 130
  6.3. Interpretation of the effects of variables on the stored energy ... 130
    6.3.1. Extent of deformation ... 130
      6.3.1.1. Stored energy and strain ... 131
      6.3.1.2. Stored energy and dislocation structure (substructure) ... 132
      6.3.1.3. The estimation of dislocation densities from stored energy, density and resistivity measurements ... 138
      6.3.1.4. Stored energy and cell structure ... 140
      6.3.1.5. Stored energy and lattice strain ... 141
      6.3.1.6. Dislocation densities from analyses of X-ray line broadening ... 142
      6.3.1.7. Stored energy and recovery ... 142
    6.3.2. Temperature of deformation ... 143
      6.3.2.1. Annealing of point defects ... 144
    6.3.3. Rate of deformation ... 146
    6.3.4. The deformation process ... 146
    6.3.5. Deformation history ... 148
      6.3.5.1. Bauschinger effect ... 148
      6.3.5.2. Work softening ... 149
    6.3.6. Cyclic deformation (fatigue) ... 149
    6.3.7. Grain structure ... 150
      6.3.7.1. Single crystals ... 151
    6.3.8. Purity ... 153
    6.3.9. The identity of the metal ... 154

6.3.10. Composition and configuration of alloys	155
6.3.10.1. Solid solutions	155
6.3.10.2. Alloys with long-range order	157
6.4. Kinetics of the release of the stored energy	160
6.4.1. Recovery	160
6.4.1.1. Energy release immediately following deformation	162
6.4.2. Recrystallization	162
6.4.3. The role of the stored energy in restoration processes	164
6.5. The dissipation of energy during deformation	165
6.6. Summary of the contributions of imperfections to the stored energy	166
6.6.1. Dislocations	167
6.6.2. Point defects	167
6.6.3. Stacking faults—twins and twin faults	167
6.6.4. Destruction of order	168
CONCLUSION	168
ACKNOWLEDGMENTS	169
REFERENCES	170
AUTHOR INDEX	179
SUBJECT INDEX	183
CONTENTS OF PREVIOUS VOLUMES	187

THE STORED ENERGY OF COLD WORK

M. B. BEVER
Dept. of Metallurgy and Materials Science, MIT, Cambridge, Mass.

D. L. HOLT
Dept. of Mechanical Engineering, University of Auckland, N.Z.

and

A. L. TITCHENER
Dept. of Chemical and Materials Engineering, University of Auckland, N.Z.

INTRODUCTION

The stored energy of cold work was the subject of a review published in this series in 1958.[1] That review covered the results of approximately fifty investigations conducted over a period of nearly six decades. Since the publication of the earlier review, the number of investigations of the stored energy of cold work has more than doubled and much progress has been made with the interpretation of the experimental data. Several surveys of specific aspects of the stored energy of cold work have also appeared.[2-5]

The early publications on the stored energy of cold work were primarily concerned with the determination of experimental values. In the mid-nineteen-fifties, however, the effects of several important variables had not yet been measured and some of the published findings were in disagreement. Since then the body of experimental data has greatly expanded. Some of the earlier discrepancies in the data have been removed, but others remain and new discrepancies have arisen. In some instances, new experimental results have made it possible to reappraise earlier values. The most significant developments have taken place in the interpretation of the stored energy, particularly regarding its dependence on variables and its relation to other property changes due to cold work. Consequently, the subject of the stored energy of cold work has become closely related to the subject of crystal defects, the theory of work hardening and the theory of restoration processes of cold-worked metals.

This review is intended to be self-contained. It deals more briefly than the earlier review,[1] however, with the description of the experimental methods of measuring the stored energy and the discussion of interpretations of only historical interest.

THE NATURE OF THE STORED ENERGY OF COLD WORK

When a metal is plastically deformed by cold working, most of the mechanical energy expended in the deformation process is converted into heat, but the remainder is stored in the metal, thereby raising its internal energy. This storing of energy is characteristic of the process of cold working, and the energy stored by the metal is an essential feature of the cold-worked state. Cold-worked metals are unstable and, under favorable conditions, undergo the restoration processes of recovery and recrystallization; during these processes stored energy is released. Thus, the stored energy of cold work is important for the process of deformation by cold working, the nature of the cold-worked state, and the restoration processes that occur in cold-worked metals.

A metal subjected to a stress deforms either elastically or plastically. The distinguishing criterion for plastic deformation is the permanent "set" present after removal of the stress. Plastic deformation is conventionally divided into hot and cold work. Cold work may be defined in terms of the effects it has on the structure of the deformed metal; it is a permanent deformation causing distortion and disarrangement of the lattice. Cold work is accompanied by work hardening, that is, an increase in resistance to further deformation. If plastic deformation is carried out under certain conditions, its effects on the structure of the metal are partly or wholly lost. These conditions include a minimum temperature and also depend on the metal, the strain, the strain rate and other variables. Deformation above this temperature under appropriate conditions is called hot work. It is in the nature of hot working that little or no energy is stored during this process.

Cyclic deformation (fatigue) is related to cold work, but in general it does not result in a permanent set. Fatigue produces structural changes not unlike those produced by cold work. The increases in the internal energy associated with these changes will be covered in this review.

Shock loading (explosive loading) is another deformation process related to cold work, but is characterized by only a slight permanent set, which furthermore is usually not reproducible. The structural changes resulting from shock loading have energy effects, which will also be discussed.

The stored energy is the change in internal energy arising from the plastic deformation. An essential characteristic of the stored energy of cold work is that it remains in the metal until it is released during a restoration process. Removal of the deforming force does not release this energy, and this distinguishes the stored energy of cold work from the macroscopic elastic strain energy.

The present knowledge of the stored energy of cold work will be reviewed under the following headings:
1. Thermodynamic aspects of deformation processes.
2. Methods of measuring the stored energy.
3. The effect of variables on the amount of the stored energy.
4. The effect of variables on the release of the stored energy.
5. Other properties measured in investigations of the stored energy.
6. Interpretation.

1. THERMODYNAMIC ASPECTS OF DEFORMATION PROCESSES

The deformation of a body can be analyzed in terms of the First Law of Thermodynamics

$$\Delta E = Q + W \tag{1.1}$$

where ΔE is the change in internal energy of the body, Q is the heat effect associated with the deformation (positive if absorbed) and W is the work (positive if done on the body). For an analysis by the First Law, a process need not be carried out under specified conditions, such as isothermally or adiabatically; in particular, thermodynamic reversibility is not a necessary condition. The First Law, therefore, can be applied to both elastic and plastic deformation.

In a deformation process the work W is the expended mechanical energy. In Parts 2 to 6, the designation W will be replaced by E_w, especially in the ratio of the stored energy to the expended energy. The heat effect Q is that which returns the body to its initial temperature. With this specification of Q, ΔE is a positive quantity in the usual processes of cold working of a metal and equals the stored energy of cold work E_s, as shown by the following relations:

$$E_s \equiv \Delta E = Q + W \qquad (1.2)$$

$$Q < 0 < W; |Q| < |W| \qquad (1.3)$$

In the special case in which the energy equivalent of work softening occurs in a secondary deformation (see 3.1.5.2), $|Q|$ is larger than $|W|$ and ΔE is negative.

Some methods of measuring the stored energy of cold work determine the change in internal energy ΔE (see section 2.1). Others determine the change in enthalpy ΔH (see section 2.2). At constant pressure, however,

$$\Delta H = \Delta E + P\Delta V \qquad (1.4)$$

where P is the hydrostatic pressure and ΔV the volume change associated with the process. Since the product $P\Delta V$ is negligibly small for solids at or near atmospheric pressure, the enthalpy change ΔH is approximately equal to the change in internal energy ΔE, which is the stored energy E_s.

1.1. The Work of Deformation

The mechanical work done on a body is given by

$$W = \sum_{1}^{n} \int_{0}^{u_i} f_i du_i \qquad (1.5)$$

where f_i ($i = 1 \ldots n$) is the instantaneous value of the ith force acting on the body and u_i is the displacement of the point of application of the force. Equation (1.5) applies to elastic and inelastic deformation. It gives the work of deforming the body, the work against friction, and the work that increases the kinetic and potential energy of the body; the last two may be neglected for deformation processes.

The work of deformation can be written

$$W = \int_V \int_0^{\bar{\varepsilon}_{ij}} \bar{\sigma}_{ij} d\bar{\varepsilon}_{ij} dV \qquad (1.6)$$

where $\bar{\sigma}_{ij}$ is the instantaneous value of the stress tensor and $d\bar{\varepsilon}_{ij}$ the infinitesimal increment of true strain; V designates the volume of the body. For uniaxial deformation eq. (1.6) becomes

$$W = \int_V \int_0^{\bar{\varepsilon}} \bar{\sigma} \, d\bar{\varepsilon} dV \qquad (1.7)$$

where $\bar{\sigma}$ is the true stress and $\bar{\varepsilon}$ the true strain. The differences should be noted between true stress ($\bar{\sigma} = f/A$) and engineering stress ($\sigma = f/A_0$) and between true strain [$d\bar{\varepsilon} = dl/l$; $\bar{\varepsilon} = \ln(l/l_0)$] and engineering strain [$d\varepsilon = dl/l_0$; $\varepsilon = (l - l_0)/l_0$] where f is the applied force, and A and l represent instantaneous and A_0 and l_0 initial values of the cross-sectional area and the length. At small strains, the differences between $\bar{\sigma}$ and σ and between $\bar{\varepsilon}$ and ε

are negligible, but at large strains they become significant. In this review normal strains are expressed as true strains. For shear strain the conventional engineering definition is adopted. For torsion the strain will be generally expressed as the nondimensional quantity nd/l, where n is the number of turns, d the diameter of the specimen and l its length. This quantity is $1/\pi$ times the maximum shear strain γ_{max} at the surface of a twisted cylindrical specimen. The latter strain will also be used.

In problems involving the energy of deformation, the evaluation of the work integral is often required. In the special case of linear elasticity, the work of uniaxial extension from an initial state of zero stress is given by the relation

$$W = \tfrac{1}{2}\sigma\varepsilon = \frac{\sigma^2}{2Y} \tag{1.8}$$

where W is the work per unit volume and Y is Young's modulus. Corresponding equations can be developed for more complex elastic stress states.

For plastic deformation a simple equation relating stress and strain is not generally available. However, the work of deformation can usually be obtained from measured force–displacement data or their equivalents; graphical integration is often necessary. In this connection the nature of the various experimental curves must be kept in mind. For example, with uniaxial extension, the areas under the curves relating (1) the force to the elongation, (2) the engineering stress to the engineering strain, and (3) the true stress to the true strain all give the correct values of the total work of deformation and the work per unit volume, as long as the extension is uniform. Beyond the onset of instability, the first two curves yield the correct value of the total work, but only an average value of the work per unit volume for the specimen as a whole. The curve relating true stress to true strain, however, may be used up to fracture to give values of the work per unit volume for any element for which the strain is known.

The work of deformation depends on the path, as shown by eq. (1.6). It is not solely a function of the final strains (shape change). As pointed out by Gubkin and Bogdanov,[6] the change in the physical state of the deformed body also depends on the path of deformation. An example is provided by the cyclic deformation of a cube in which it is deformed into a rectangular prism and then restored to its original cubic shape. The final strains are zero, but the work of deformation and the changes in some physical properties are not.

The path of deformation in industrial processes such as wire drawing, rolling and swaging is not as simple as in tension or compression nor is the path of deformation the same for every element of the deforming material. Consequently the work done to produce a given shape change must be different. It is known to be larger if the change is produced by wire drawing, rolling and swaging than if it is produced by tension or compression. The amount by which it is larger is the so-called redundant work W_r. In addition, the work in industrial deformation processes is increased by the friction work W_f. Often neither W_f nor W_r is known and the actual work expended W may then be related to the ideal work W_i by a factor η, the efficiency of deformation: $\eta W = W_i$. Redundant work increases the amount of energy stored for a given shape change. Friction work does not contribute to the stored energy. It may reduce it because it raises the temperature of the material during working and therefore promotes recovery.

Most published values of the stored energy of cold work have been reported as a function of the largest normal or shear strains. Each of these strains is a suitable parameter for a given process of deformation. Neither, however, adequately defines the extent of deformation if different deformation processes are to be compared. The stored energy is likely to be a function of all dimensional changes and the deformation path. An additional difficulty in using the largest strain as a measure of the extent of deformation is that in many processes, for example torsion, the strain is not uniform throughout the specimen.

The expended energy E_w (which is equal to the mechanical work W) is often used in this review as a measure of the extent of deformation. This facilitates the comparison between different processes. The problem involved in correlating the energy stored in different processes is not completely eliminated, however, since the energies stored by specimens subjected to different shape changes with the same expenditure of energy, all other constraints being the same, will not necessarily be equal. A disadvantage of the parameter is that it includes the work done against friction in such processes as wire drawing and drilling. Also, the expended energy has not been reported in all investigations.

1.2. Thermal Effects of Deformation

1.2.1. *Elastic Deformation*

In an elastic deformation under adiabatic conditions, the temperature of the deforming specimen changes. In a uniaxial elastic deformation of a bar of unit cross-section, the adiabatic temperature change, according to an equation proposed in 1851 by W. Thomson (Lord Kelvin)[7,8] is

$$\Delta T_s = -\frac{\alpha T \Delta \sigma_s}{c_\sigma} \tag{1.9}$$

In this equation, α is the coefficient of linear thermal expansion, T the absolute temperature, $\Delta \sigma_s$ the isentropic change in stress, and c_σ the heat capacity per unit volume at constant stress.

It follows from eq. (1.9) that, in normal materials having a positive coefficient of thermal expansion, the temperature decreases during adiabatic elastic extension and increases during adiabatic elastic compression. This temperature change, which is a manifestation of the thermoelastic effect, is usually smaller than 0.2°C in metals. It has been measured by numerous investigators since its first observation in 1830.[9] The thermoelastic temperature change must be allowed for in single-step methods of measuring the stored energy (see section 2.1), in which the heat produced by the deformation is obtained from the change in temperature of the specimen.

If an elastic deformation is to be carried out isothermally rather than adiabatically, heat must flow into or out of the specimen to offset the thermoelastic heat effect. In an isothermal cycle of perfect elastic loading and unloading, the heat absorption (or evolution) during the first step cancels the heat evolution (or absorption) during the second step. In an adiabatic cycle, the temperature changes cancel correspondingly. In this sense, the thermoelastic effect is conservative rather than dissipative.

1.2.2. *Plastic Deformation*

If a metal is plastically deformed under adiabatic conditions, its temperature rises; if the plastic deformation is carried out isothermally, heat is evolved. These effects are characteristic of plastic deformation and, in contrast to the thermoelastic effect, have the same sign for such complementary deformation processes as extension and compression.

Tammann and Warrentrup[10] investigated the temperature changes associated with elastic and plastic deformation. They found that low-carbon steels in tension obeyed Thomson's equation up to the yield point, where a sharp reversal in the direction of the temperature change indicated the onset of plastic deformation. In unannealed copper and nickel, the reversal was more gradual and no yield point could be assigned to them on the basis of temperature measurements.

The ideal plastic material by definition deforms with no work hardening. For the plastic deformation of such a material, as Bridgman[11] has pointed out, $\Delta E = 0$ and $Q = -W$. i.e., the whole of the work is converted into heat. No actual metal is ideally plastic. Mild steel approximates ideally plastic stress–strain behavior at the lower yield point, but it is unlikely that the associated rate of energy storage is unusually low. "Easy glide" of a single crystal is another approximation to ideal plastic stress–strain behavior: the associated energy effects will be dealt with in 3.2.1.1, 4.2.1.1 and 6.3.7.1.

1.3. Changes in Thermodynamic Functions with Deformation

1.3.1. Internal Energy

Elastic and plastic deformation change the internal energy of a metal. In an isothermal elastic deformation, there are two contributions to this energy change, namely, the elastic strain energy and the thermoelastic heat effect. In a uniaxial deformation obeying Hooke's Law and starting from an unstressed state, the increase at constant temperature in internal energy per unit volume is

$$\Delta E = \frac{\sigma^2}{2Y} + \alpha \sigma T \qquad (1.10)$$

Calculations show that at ordinary temperatures, the second term is much larger than the first for iron, copper, aluminum and magnesium. Thus, the internal energy of these metals, and probably of others, decreases in compression. Elastic deformation, however, does not produce "stored energy" in the sense defined in the Introduction, because removal of the deforming forces results in an immediate relaxation, and the internal energy returns to its original value.

Plastic deformation by cold work may require the expenditure of much larger amounts of energy than are encountered in elastic deformation, but usually only a small fraction of this energy is retained. The increase in internal energy caused by cold work is the stored energy E_s.

1.3.2. Free Energy

Cold work changes the free energy as well as the internal energy of a deformed metal. The change in free energy is the true measure of the thermodynamic instability resulting from the deformation.

At first sight it may appear possible to determine the Gibbs free energy change associated with cold work by measurements in an electrolytic cell according to the equation

$$\Delta G = -n\mathscr{F}\mathscr{E} \qquad (1.11)$$

where n is the valence of the active ion, $\mathscr{F}$ Faraday's constant, and $\mathscr{E}$ the open-circuit potential of the cell. The cell

$$\text{Ag(cryst)/AgNO}_3\text{(aq)/Ag(cold-worked)}$$

may be considered as an example. A current passing from the right to the left dissolves cold-worked silver and deposits normal crystalline silver at the left-hand electrode; a current passing in the opposite direction dissolves normal silver, but does not deposit cold-worked silver at the right-hand electrode. The cell, therefore, is not reversible and eq. (1.11) cannot be used. Even if an electrolytic method could be used, it would encounter other difficulties. The stored energy is not distributed uniformly throughout a piece of cold-worked metal, but resides in local regions; on a surface having such an energy distribution, local cell action would occur. Also, whatever e.m.f. measurements were to be made would refer to the surface of the electrode, which, because of partial relaxation, is not representative of the interior.

Vapor pressure measurements may seem to offer a possibility for the direct determination of the free energy change associated with cold work. Apart from likely experimental difficulties, however, the rather high temperatures required would result at least in partial recovery. Also, as with the e.m.f. method, any measurement would apply only to the surface.

The Gibbs free energy change associated with cold work may be evaluated from the following equation if the required data are available

$$\Delta G = \Delta H - T\Delta S \tag{1.12}$$

Since the enthalpy and internal energy changes due to cold work are approximately equal, we can write

$$\Delta G \approx \Delta E - T\Delta S = \Delta F \tag{1.13}$$

where ΔF is the Helmholtz free energy change. Enthalpy or internal energy changes due to cold work are measured by the experimental methods to be described in Part 2. A knowledge of the entropy effects is also required if the free energy change is to be evaluated.

1.3.3. *Entropy*

The entropy change of a body undergoing a physical change can often be evaluated from the equation

$$\Delta S = \frac{Q_{\text{rev}}}{T} \tag{1.14}$$

where Q_{rev} is the heat effect for a reversible path between the initial and final states. Since no reversible path can be conceived between the annealed and the cold-worked states of a metal, the entropy change associated with cold working cannot be evaluated by eq. (1.14).

The change in entropy is also given by

$$\Delta S = k \ln \frac{w_2}{w_1} \tag{1.15}$$

where w_1 and w_2 are the possible numbers of microstates that the system may assume in the initial and final states and k is Boltzmann's constant. The total entropy change is the sum of changes due to configurational, vibrational, rotational, electronic, and other effects.

In general, only the configurational and vibrational contributions are significant in the cold working of metals and the resulting entropy change is

$$\Delta S = \Delta S_{\text{config}} + \Delta S_{\text{vibr}} \qquad (1.16)$$

Apart from the idealized case of simple translation in a perfect single crystal, plastic deformation invariably increases the structural disorder, and thus raises the configurational entropy. The change in vibrational entropy, however, may be positive or negative. It is possible, at least in principle, to estimate the contributions to the entropy made by various types of imperfections present in cold-worked metals, as discussed in section 1.4.

1.3.4. *Heat Capacity*

In anisothermal annealing methods (see subsection 2.2.1), the stored energy of cold work is determined as the difference between the amounts of heat required to raise a cold-worked and an annealed specimen through the same temperature interval. This difference does not result from a difference in the true heat capacities of the cold-worked and annealed specimens; heat capacities apply to reversible changes, but an irreversible change takes place in the cold-worked specimen during heating. The true heat capacity of a cold-worked metal can be measured only below the lowest temperature at which stored energy is released. In general this is the temperature at which the metal begins to recover at an appreciable rate.

The possibility that cold work might affect the true heat capacity has long been recognized.[12] Early investigations of this problem were inconclusive.[13] Among investigators of the stored energy, Suzuki[14] referred to the possibility of an alteration in the heat capacity due to changing the modes of atomic vibration by means of cold work. Maier and Anderson[15] measured the heat capacity of annealed and drawn wires at temperatures below the temperature of drawing, but any change due to cold work fell below the sensitivity of their method. Martin[16] found that heavy deformation of 99.999% copper below room temperature raised the heat capacity above the value of annealed copper by about 0.15% over the temperature range 20° to 300°K. In an investigation by Ahlers[17] the heat capacity of a cold-worked specimen of copper below 6°K was 0.7% larger than that of an annealed specimen; the difference decreased to 0.3% between 6° and 9°K and remained at that value up to 21°K. Pervakov and Khotkevich[18] found that the heat capacity of specimens of gold, silver and copper cold-worked at room temperature was slightly larger than that of annealed specimens at temperatures below four-tenths of the Debye temperature.

1.4. The Energies and Entropies of Imperfections

Cold work generates and rearranges crystal imperfections. In particular, it increases the density of dislocations. Point defects may also be present in cold-worked metals in numbers greatly in excess of the equilibrium concentrations. Other imperfections which may originate during cold work are stacking faults and twins.

1.4.1. *Dislocations*

The strain energy of one dislocation E (its self-energy) may be estimated[19] from its strain field as

$$E \approx \frac{\mu b^2}{4\pi K} \ln \frac{R}{r_0} \tag{1.17}$$

where μ is the shear modulus, b is the Burgers vector of the dislocation and R and r_0 are the outer and inner radii of the strain field. K is a coefficient having the values 1 for a screw dislocation, $(1 - v)$ for an edge dislocation (where v is Poisson's ratio) and intermediate values for a mixed dislocation. The corrections for crystalline anisotropy appear to be small.[20] It may be shown that r_0 is of the order of the Burgers vector and that the energy of the core within the radius r_0 is only a small fraction of E.[19-21] If plausible values of R are inserted into eq. (1.17), the energy is found to lie within the range 4–9 eV per atom plane pierced by the dislocation.[19,22] The energy of a screw dislocation is about half that of an edge dislocation. The elastic strain energy is a free energy.

The configurational and vibrational entropy effects of a dislocation were estimated by Cottrell[19] to be not larger than 10^{-4} eV per °K per atom plane. Hence, at ordinary and low temperatures, the temperature-entropy product may be neglected and the internal energy change of a dislocation may be set equal to the free energy change.

When more than one dislocation is present in a crystal, the total energy is the sum of the self-energies and the interaction energies. The interaction energies may be positive or negative and may depend critically on the arrangement of the dislocations. For example, Li showed that a small change in the angle between two crossed grids of screw dislocations produces a large change in the strain energy.[23] The energies of assemblies of dislocations will be discussed in subsection 6.3.1.

1.4.2. *Point Defects*

Point defects cause energy and entropy changes. In a crystal of N atoms containing n point defects, the configurational entropy increase per defect is

$$\Delta S_{\text{config}} \approx -k \ln (n/N) \tag{1.18}$$

The vibrational entropy change, according to Vineyard and Dienes,[24] depends only on the lattice frequencies v of the perfect crystal and v' of the imperfect crystal, and needs no allowance for the temperature dependence of the enthalpy of formation. The vibrational entropy change per point defect is therefore

$$\Delta S_{\text{vibr}} = k \sum_i \ln \frac{v_i}{v_i'} \tag{1.19}$$

where the values v_i' are the frequencies of the vibrational modes of a crystal containing one additional vacancy but the same number of atoms as a crystal in which the vibrational modes are $v_1, v_2, \ldots v_n$ and the sum is to be extended over all independent vibrational modes. Equation (1.19) holds only if $hv_i \ll kT$.

A crystal structure at equilibrium contains a concentration $x_v = n/N$ of vacancies given by

$$x_v \approx \exp(-G'_{f,v}/kT) \tag{1.20}$$

where $G'_{f,v}$ is the free energy of formation exclusive of the configurational entropy term according to the equation

$$G'_{f,v} = E_{f,v} - TS_{f,v(\text{vibr})} \tag{1.21}$$

In metals the entropy of vibration of a vacancy appears to be positive, but small; it is probably about $(1.5 \pm 0.5)\,k$ in face-centered cubic metals.[25]

Thus

$$G'_{f,v} \approx E_{f,v} \qquad (1.22)$$

and

$$x_v \approx \exp(-E_{f,v}/kT) \qquad (1.23)$$

Vineyard and Dienes[24] have suggested that the entropy of vibration of interstitials in metals may be negative. No values are known.

The energy of formation of vacancies has been estimated.[20] It has also been measured for various metals. Averages of experimental values are shown for selected metals in column 4 of Table 1.1. This table includes measured values of the energies of migration of vacancies $E_{m,v}$ and of the activation energies of self-diffusion E_D. As may be seen, the sum of $E_{f,v}$, the energy of formation of vacancies, and $E_{m,v}$ is always approximately equal to E_D. This is to be expected if diffusion takes place by a vacancy mechanism. Therefore, in cases in which one of the three quantities is unknown, it may be deduced from the other two.

Table 1.1 lists measured values of the energy of migration of interstitials, $E_{m,i}$. The energy of formation of an interstitial has been estimated for copper and probably lies in the range 3–5 eV.[20] No experimental values are available for interstitials in metals. Simmons and Balluffi[25] have shown that the concentration of thermal interstitials is negligible. Appreciable numbers of interstitials may be present after cold work at low temperature.

The energies of dislocations and point defects will be discussed in the interpretation of the stored energy in Part 6. The energies of other imperfections such as stacking faults, and the energies associated with such special mechanisms as the destruction of short-range order will be dealt with in Part 6.

1.5. Irreversible Thermodynamics and Plastic Deformation

Elastic deformation as an ideally reversible process can be analyzed by classical thermodynamics. Classical thermodynamics can also be employed, although in a more restricted way, for the analysis of plastic deformation by cold work. Specifically, the First Law of Thermodynamics can be applied. Consequently the change in internal energy, i.e.; the stored energy of cold work, is a determinate quantity [see eq. (1.2)]. Also, it follows from the Second Law that in plastic deformation the change in the Helmholtz free energy must be equal to or less than the expended work. Plastic deformation has non-equilibrium characteristics, however, which cannot be treated by classical thermodynamics; their analysis makes recourse to irreversible thermodynamics necessary.

Irreversible thermodynamics was first developed for steady-state processes.[26, 27] Since plastic deformation by cold work is not a steady-state process, irreversible thermodynamics in its original form and particularly Onsager's relations do not apply to it.

Plastic deformation is an extreme case of irreversibility. Not only is it not reversible, but no reversible path for producing the cold-worked state can be conceived. Bridgman aptly made this point by describing the cold-worked state as "an island in a sea of irreversibility".[11]

Kestin considered a cold-worked body as an aggregate of elastic-plastic domains that deform in an ideally plastic manner under the action of sufficiently large stresses.[28]

TABLE 1.1
Energies of Interstitials and Vacancies in Metals

	Interstitials	Vacancies		
	$E_{m,i}$	$E_{m,v}$	$E_{f,v}$	E_D
Al	0.1	0.65	0.8	1.45
Ag	0.1	0.9	1.1	1.9
Au	0.08	0.95	1.0	1.8
Cu	0.11	≈0.8	1.15	2.0_5
α-Fe	0.2	1.2	—	2.8
Mo	—	1.2_5	—	4.1
Ni	0.1	1.0_5	—	2.8_5
Pb	—	—	0.6	1.0_5
Sn	—	0.6_5	0.5	1.0_5
W	0.5	1.7	—	5.2

Source: Friedel.[20]
See text for definition of symbols.

According to his analysis, energy is stored elastically during the deformation by a field of locked-in microstresses. Rubin, however, pointed out some deficiencies of a simple model of this kind.[29] In particular, work hardening according to the model would be accompanied by the release of "stored strain energy". He argued that "this correspondence between hardening and release of stored energy ... is in contradiction with the behavior of real materials whose hardening is generally associated with an increase in the residual strain energy".

Recent developments of irreversible thermodynamics have introduced internal state variables as a means of dealing with processes of a more pronounced irreversibility than that characterizing steady-state processes. Bridgman[11] had made some early use of this concept; the analysis was developed by Meixner and Reik,[30] Meixner,[31] and particularly Coleman and Gurtin.[32] Kestin and Rice,[33] Chu,[34] and Kratochvil and Dillon[35] have used internal state variables in analyzing plastic deformation.

Internal state variables can represent various quantities; in the case of plastic deformation by cold work they primarily refer to dislocation parameters, such as the dislocation density and higher moments of the dislocation distribution. The internal energy change associated with plastic deformation by cold work, i.e., the stored energy of cold work, may be a useful quantity in this connection because it represents an integrated effect of changes in dislocation parameters. This integrated effect would be most readily applicable to pure metals. Solid solutions, especially ordered solutions, would introduce complications and additional internal variables; order may represent an internal variable, which changes during deformation.

Kestin and Rice,[33] with the mildly restrictive assumption that the rate of change of each internal variable is governed by its conjugate force showed that the plastic strain rate $\dot{\varepsilon}^p_{ij}$ can be expressed as

$$\dot{\varepsilon}^p_{ij}(\sigma, \xi, T) = \frac{\partial \Omega(\sigma, \xi, T)}{\partial \sigma_{ij}} \qquad (1.24)$$

where σ is the stress tensor, ξ is the set of internal variables and Ω the plastic potential. The potential has been used in continuum mechanics for some time[36-36b] but Kestin and

Rice have provided a thermodynamic basis for it. By this concept, thermodynamics can link the microscopic and macroscopic aspects of plastic deformation.

2. METHODS OF MEASURING THE STORED ENERGY

The stored energy of cold work is a comparatively small quantity. The heats of fusion of metals are typically several thousand calories per gram-atom and the heats of polymorphic transformation several hundred. In most polycrystalline metals deformed at room temperature, the stored energy does not exceed a few tens of calories per gram-atom even at large strains; in solid solution alloys it rarely exceeds 200 cal/gram-atom. The energy stored by crystals at small strains is typically less than 5 cal/gram-atom. Moreover, techniques capable of measuring energies at least one order of magnitude smaller than the maximum values are desirable for the investigation of the effect of variables on the stored energy. As a result the measurement of stored energies presents serious problems. An understanding of the difficulties and possible errors inherent in the different methods is necessary for an evaluation of published results.

Leach[37] has discussed the principles of calorimetry and has surveyed procedures and equipment for calorimetry applicable to the solid state, including stored energy measurements. Williams[5] reviewed in detail the calorimetric methods used in stored energy measurements.

The classification of the methods of measuring the stored energy adopted in the earlier review[1] is retained here. The methods are:

(1) Single-step methods: all measurements are made during the deformation; methods in which the specimen is placed in a calorimeter immediately after deformation are included in this class.

(2) Two-step methods: the deformation is carried out first and the stored energy is measured at a later time.

The broad features of single-step and two-step methods will be discussed in sections 2.1 and 2.2. Column 3 of Table 3.1 identifies the methods used in published investigations.

2.1. SINGLE-STEP METHODS

All single-step methods are based on a direct application of the First Law of Thermodynamics. The stored energy is obtained as the difference between the work done on the body and the heat evolved by it.

The total work done on the body is usually determined from force-displacement data. Other techniques have also been used. For example, the work has been determined from the loss of potential energy of pendulum hammers.[38, 39] The total work of deformation includes the elastic energy of the deformed specimen. It may also include work done on the straining device and against friction between the straining device and the specimen. The work done against friction, which may be small as in compression or large as in wire drawing, appears as heat. If this heat is not included in the total heat measured, the work term must be corrected for it.

The heat produced by the deformation is usually obtained from the measured increase in temperature of the specimen and its heat capacity. The heat has also been measured by dropping the specimen into a calorimeter immediately after deformation[40] or by carrying out the deformation within a calorimeter.[15, 38, 41–45] With all single-step methods, the measured heat must be increased by the amount of heat lost during deformation. This loss may be caused by conduction from the specimen to the straining device and by radiation and convection to the surrounding environment. In methods in which the increase in temperature of the specimen is measured, allowance also has to be made for the thermoelastic effect (see subsection 1.2.1).

If full corrections are not made to the work and heat, the value of the work is too large and, except for the correction for the thermoelastic effect in compression of a material with a positive coefficient of thermal expansion, the value of the heat is too small. The calculated values of the stored energy are thus too large. Since work and heat are measured by different techniques, systematic errors in one of the quantities do not cancel systematic errors in the other except fortuitously. By contrast, systematic errors in the calorimetry of two-step methods of measuring the stored energy tend to cancel when differences are taken.

The magnitudes of the work and heat do not usually differ by more than about 10% and may be within 1% of each other. Success in the use of single-step methods, therefore, depends on measuring these two quantities to a high degree of absolute accuracy. High precision of measurement (or reproducibility) is not sufficient.

A brief analysis serves to illustrate the critical dependence of the stored energy on the accuracy of the measurements of the work and heat and of the corrections applied to them. If the true value of the work E_w is assumed to be smaller than the measured value E'_w by a small amount jE_w and the true value Q to be larger than the measured value Q' by a small amount kQ, the difference between the experimental value E'_s and the true value E_s of the stored energy is given approximately by

$$E'_s - E_s = jE_w + kQ \approx (j + k) E'_w \qquad (2.1)$$

The corresponding difference between the ratios of stored to expended energy is given approximately by

$$\frac{E'_s}{E'_w} - \frac{E_s}{E_w} \approx (j + k) \left[1 - \frac{E'_s}{E'_w} \right] \qquad (2.2)$$

In Fig. 1 the full line is a hypothetical curve of the energy E_s stored in copper deformed by

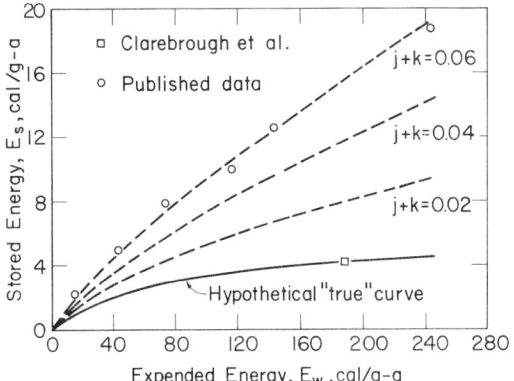

FIG. 1. Effect of systematic errors in the values of heat and work on typical published values obtained by a single-step method of measuring the stored energy. The single point on the hypothetical "true" curve is from Clarebrough et al.[46]

compression against the expended energy E_w. No specific functional relation between E_s and E_w is assumed, but the curve is drawn through a single value of the energy obtained by Clarebrough et al.[46] by an anisothermal annealing method. The curve may now be considered to represent values of E_s obtained from "true" values of E_w and Q measured by a single-step method. The dashed curves show the effects of combined errors of 2, 4, and 6% in E_w and Q ($j + k = 0.02, 0.04, 0.06$). The points lying close to the uppermost curve in Fig. 1 represent a set of published data obtained by a single-step method of measuring the stored energy in deformed copper. The figure shows that relatively small systematic errors in measuring E_w and Q cause large errors in E_s as well as a significant change in the shape of the curve. This change is even more marked when the stored energy is plotted against strain; in such a plot, the uppermost curve becomes a nearly straight line with the published values close to it, while the shape of the "true" curve changes little.

Single-step methods have been applied to tension, compression, torsion, drilling, wire drawing, cyclic straining and deformation by impact (see Table 3.1). A general limitation of these methods is that they do not furnish information on the release of the stored energy during restoration processes. Some of them, however, are uniquely suitable for following any release of energy that may occur immediately after deformation.[38, 39, 43]

Single-step methods were used in the earliest investigations of the stored energy.[47–49] They predominated numerically until about 1950, but were little used in the next decade. In the nineteen-sixties, major developments in single-step methods for measuring the stored energy were made by Williams,[38, 43, 50] Erdmann and Jahoda[44] and Wolfenden and Appleton.[45] Because single-step methods used before 1958 were described in the first review,[1] we shall consider here only developments since that time.

In the "impact calorimeter" which Williams used for several investigations,[38, 51–53] the specimen was compressed by two pendulum hammers. He calculated the work from the loss of energy of the hammers and the heat effect from the rise in temperature of the specimen. He gave the precision as $\pm 0.1\%$ of the supplied energy and the accuracy of the stored energy values as $\pm 10\%$.[5, 38]

In other investigations, Williams used a "liquid-gas film calorimeter".[43, 54] In this calorimeter, the specimen was deformed in tension and the work was determined from the force–displacement data. The heat effect was obtained from the volume of gas produced by the evaporation of a liquid film coating the specimen. Williams gave the precision of the stored energy measurements as about 5%, which he believed also to be the accuracy.

For still another investigation, Williams built an apparatus to measure the stored energy in single crystals in tension.[50] The heat effect was measured by the rise in temperature of the specimen. The instrumentation was capable of considerably higher precision than that of earlier investigators who used the same general method, but values of the precision and accuracy of the measurements are not known. Wolfenden[55–60] also conducted some of his investigations in this equipment. A similar technique was used by Dillon,[61, 64] Kunin,[65, 66] Kunin, Kunin, Grishkevich and Korenchenko,[67] Nakada,[68] Halford,[69] Ham,[39] Panin and Milevskaya,[70] Panin, Sukhovarov and Dudarev[71] and Shermergor.[72]

In apparatus used by Erdmann and Jahoda for measuring the energy stored in wires deformed at 4.2°K, the work was obtained from force-displacement data and the heat effect from the temperature rise of the specimen.[44] The accuracy and precision of the measurements were not stated. With this apparatus Erdmann and Jahoda could also measure the electrical and thermal conductivity and the thermal diffusivity.

Wolfenden and Appleton developed a phase-change calorimeter.[45] The specimen was deformed under liquid nitrogen and the heat effect was obtained from the change in the rate of flow of nitrogen gas. They estimated the accuracy of the stored energy values as $\pm 30\%$ at the highest strains.

2.2. Two-step Methods

In two-step methods the stored energy is found by comparing the thermal behavior of the cold-worked specimen with that of a standard specimen in a suitable process. In the case of metallic elements and some alloys, the standard specimen is in the annealed condition; standard specimens of ordered alloys require special thermal treatments. All two-step methods depend on converting the cold-worked and standard specimens to identical final states. They do not require the measurement of the energy expended during deformation; this quantity has not been reported in all investigations in which a two-step method was used.

Two-step methods may be classified in the following manner:
(1) Annealing methods, which may be further subdivided into:
  (a) Anisothermal annealing, in which the temperature of the specimen is increased continuously, usually at a constant rate, and the effect of the energy release is observed as a function of temperature.
  (b) Isothermal annealing, in which the specimen is placed in a constant-temperature calorimeter and the effect of the energy release is observed as a function of time at a selected annealing temperature.
(2) Reaction methods, in which cold-worked and standard specimens react with a working substance in a calorimeter, and the energy is obtained as the difference between the heats of reaction.

Two-step methods determine the enthalpy difference between the cold-worked and the standard state; single-step methods determine the difference in internal energy. Two-step methods do not measure the stored energy that may be released immediately after deformation.

2.2.1. Anisothermal Annealing Methods

In anisothermal annealing methods the stored energy is determined from the difference in the heat required to raise the temperature of a cold-worked and a standard specimen. This difference is due to the release of stored energy by the cold-worked specimen.

The most successful anisothermal annealing methods have made use of differential techniques in which the test specimen and the standard specimen are placed in identical environments in an externally heated calorimeter chamber. In one of these methods the difference in temperature between the test specimen and the standard specimen is measured while both are absorbing heat from the chamber.[73, 74] In a second method auxiliary heaters are placed within the two specimens and a wattmeter measures the difference in power required to raise their temperature at equal rates.[75] J. L. White and Koyama[74] have designated these two methods "differential thermal analysis" and "differential power analysis". In both methods corrections necessary to compensate for small accidental differences between the thermal environments of the two specimens can be made by establishing a base line; this is usually done by a second heating of the test and standard specimens.

Differential thermal analysis does not yield quantitative results directly unless the difference between the temperature of the specimen and its surroundings is measured during the heating. In the earliest determination of the stored energy by differential thermal analysis, Sato[73] measured only the temperature difference between the test specimen and the standard specimen. To calibrate his temperature-time curves he measured the energy stored in a cold-worked 70Cu–30Zn alloy in a separate high-temperature drop calorimeter and equated this quantity to the area under the temperature–time curve obtained by differential thermal analysis of the same alloy deformed to the same strain.

Anisothermal methods based on differential techniques have the advantage that they eliminate some systematic errors. Anisothermal methods that do not employ differential techniques[14, 76–86] do not have this advantage and have generally proved inferior to those that do. In anisothermal methods of any kind, an initial period is usually required to establish steady-state conditions. This need not introduce a difficulty, however, since the thermal transients decay to negligible levels before the release of energy begins if the heating is started at a sufficiently low temperature. In this respect, anisothermal methods have an advantage over isothermal methods (subsection 2.2.2), in which thermal transients cannot be avoided during the initial period. Anisothermal methods in common with isothermal methods reveal the kinetics of the release of the energy, but the analysis of these kinetics is more straightforward for the latter. Anisothermal methods have been successfully applied to measuring the stored energy released below room temperature.[87–92]

The best anisothermal annealing methods based on differential techniques have reached high levels of precision. In this respect they are superior to single-step methods and reaction methods and are probably only slightly inferior to the best isothermal annealing methods. They have higher absolute accuracy than any other method in current use.

An anisothermal annealing method was first used by Krivobok.[93] The details of his experiments as well as of other investigations made before 1958 were summarized in the first review.[1] In the intervening time, developments in anisothermal annealing methods have been made by Sizmann and Wenzl[94] and Wenzl,[95] Popov, Tikhomina, Skuratov and Kalinina,[82] Hertsriken, Larikov and Slyusar,[96] White and Koyama,[74] van den Beukel,[91] Taoka, Suzuki, Yoshikawa and Okamoto,[97] Halford,[69] Kovacs,[83] Lugscheider,[98] Lugscheider and Wildhack[92] and Ebel and Lugscheider.[99] Chin and Grant[84, 85] made stored energy measurements in a constant temperature-gradient calorimeter built by Misra, Howlett and Bever.[86]

Sizmann and Wenzl[94] did not use a differential method. They measured the difference in temperature between the specimen and a large copper block surrounding the specimen as the block was heated at a constant rate of about 1.5°C/min. They obtained a base curve by reheating after the first annealing. The precision of measurement was stated as 10% for a value of the stored energy of 1.4 cal/gram-atom of copper. Wenzl[94a] reported that subsequent improvements increased the precision by a factor of twenty.

Popov et al.[82] also did not use a differential method. They measured the difference in temperature between pieces of silver in contact with the specimen and an enclosing massive silver shell. The shell was heated inside a tubular electric furnace. They obtained a base curve by a second heating. They gave the accuracy as 50%.

Hertsriken et al.[96] used differential power analysis. The heating rate was 6°C/min. The calorimeter was calibrated by measuring the heat of fusion of tin.

White and Koyama[74] developed equipment for differential thermal calorimetry. They placed the cold-worked and annealed specimens inside a copper block which served as the calorimeter chamber. The temperatures of the specimens and calorimeter were measured continuously during heating at a constant rate. In a second heating, the correction for any differences in the thermal coupling of the two specimens to the copper block was determined. The technique for analyzing the data was similar to that of Henderson and Koehler.[88] The accuracy of the method was checked by measuring the heat of fusion of lead. The precision was estimated to be ± 0.4 cal/gram-atom of copper. The equipment could be used up to temperatures of at least 700°C.

A calorimeter employing differential power analysis that could be used at temperatures as low as 93°K was described by van den Beukel.[91] The cold-worked specimen and a standard were placed adjacent to each other, but not in contact, in a cavity of a solid block. The temperature of the block was increased at a constant rate, and the rate at which heat flowed from the block to each specimen was measured by a bank of thermocouples that separated the specimens from the block. The apparatus was calibrated by redetermining the heat capacity of a material that was not specified in the paper. The calibration was verified by measuring the heat of fusion of mercury. The average accuracy of the stored energy measurements was stated to be ± 0.03 cal/gram.

The calorimeter used by Halford[69] was similar to that used by Clarebrough et al.[75] The temperature of the calorimeter was increased at a constant rate by external heating. The specimen and the standard were heated internally to maintain them at the same temperature as that of the calorimeter. The difference in the power supplied to the specimen and the standard was measured by a sensitive differential wattmeter. Halford did not state whether a second heating was used to fix a base line or whether a check was made of the accuracy. He gave the precision as ± 0.05 cal/gram of copper.

Taoka et al.[97] also modeled their apparatus after that of Clarebrough et al. They did not report whether a second heating was used to determine the base line and did not state the accuracy and precision of their measurements.

Kovacs[83] used anisothermal annealing but not the differential method. He measured the power required to maintain the specimen at the temperature of a copper block surrounding it. The block was placed inside a furnace heated at a constant rate; a base line was obtained by a reheating.

Lugscheider[98] and Lugscheider et al.[92, 99] used differential power analysis. The heating rate was 15°C/min. The calorimeter could operate at a temperature as low as 80°K.

In the technique employed by Chin and Grant,[84, 85] the time required for heating the specimen through a known temperature interval under a known temperature gradient was compared with the time required for heating a standard specimen of known heat capacity through the same temperature interval under the same temperature gradient.

2.2.2 Isothermal Annealing Methods

In isothermal annealing methods the difference in temperature between a test specimen and a surrounding isothermal jacket of large heat capacity is measured as a function of time. The specimen and the jacket are initially as nearly as possible at the same temperature. The thermal resistance between the specimen and the jacket is large and reproducible. As the specimen releases stored energy, its temperature rises above that of the jacket; the difference between the two temperatures is a function of the rate of release of energy. An isothermal calorimeter is calibrated by releasing energy at a known rate within the working chamber.

The main advantage of isothermal calorimetry is that it reveals the kinetics of the energy release under conditions permitting simple analysis. Its main weakness is that an unknown amount of energy is lost during the initial period in which the temperature of the test specimen is being raised to that of the calorimeter. This loss becomes important if an appreciable amount of energy is released at temperatures below the temperature of isothermal annealing. Even when the specimen has been preheated externally as nearly as possible to the temperature of the calorimeter and is then inserted in it, there is a period of transient heat flow. The amount of energy released in this period, which may be from 15 min to 1 h, can be estimated only by backward extrapolation of the power–time curves to zero time. Consequently the amount of energy released in this period is uncertain. The losses during the heating up and the transient periods make the accuracy of the isothermal methods somewhat lower than that of the best anisothermal methods.

Isothermal calorimetry for measuring the stored energy was first used by Borelius and co-workers[100–103] and has since been used by others.[37, 104–112] Equipment developed since 1958 was described by Bailey and Hirsch,[107, 108] Bell and Krisement[109, 110] and Bell.[111, 112]

Bailey and Hirsch[107, 108] determined the energy released during the initial period by extrapolating the release curve back to zero time according to the relation $dQ/dt = A/(t + t_0)$ where A and t_0 are constants. They estimated the accuracy of the total measured stored energy to be 10%.

Bell and Krisement[109, 110] and Bell[111, 112] used a differential calorimeter in which the cold-worked specimen and the standard specimen placed symmetrically were heated together. The calorimeter could be operated anisothermally, but was primarily used as an isothermal calorimeter. The temperature range extended from approximately 25° to 350°C. No error limits were stated.

2.2.3. Reaction Methods

In reaction methods the stored energy is obtained as the difference between the heat effects when a cold-worked and a standard specimen react with another substance to form products having identical final states. In early applications[113–116] the reactions selected had large heat effects and the errors were large.

A reaction method that has been used by Bever and co-workers takes advantage of the small magnitude of the heats of solution of metals in molten metals.[37, 117, 118] Suitable combinations of solvent and solute metals make the heat effects associated with dissolution as close to zero as desired. The errors inherent in obtaining the stored energy as a small difference between two large quantities are thus greatly reduced. For good accuracy the amount of heat transferred during the measurement should be small and therefore the rate of solution should be high. Systematic errors in the heat effects measured by reaction methods largely cancel and the absolute accuracy of the heat effects is therefore close to their precision. By contrast, systematic errors in the measurement of heat and work in single-step methods cancel only fortuitously so that the accuracy is in general not as high as the precision.

The reaction method using metal solution calorimetry is accurate to approximately ± 2 cal/gram-atom under best conditions. The method has been applied especially to alloys, partly because the best combination of solute and solvent is obtained in this way, partly because the amount of energy stored in alloys is higher than the amount stored in the pure component metals. The method is well suited to measuring the energy stored in metals deformed below room temperature.

3. THE EFFECT OF VARIABLES ON THE AMOUNT OF THE STORED ENERGY

The amount of the stored energy depends on a number of variables. These variables can be divided into two classes: those related to the process of deformation and those related to the material being deformed. Among the former are the extent, rate and temperature of deformation; among the latter are the identity, purity and grain structure of the material.

In Part 3 we shall deal with the effects of the variables on the amount of the stored energy and on the ratio of the stored to the expended energy E_s/E_w. We shall discuss the dependence of the kinetics of the release of the stored energy on variables in Part 4.

Table 3.1 summarizes the published values of the stored energy and indicates their dependence on variables. The table includes values of the ratio E_s/E_w whenever these have been reported; it also mentions other properties measured in each investigation.

3.1. Variables Related to the Deformation Process

3.1.1. The Extent of Deformation

The stored energy has been investigated as a function of the extent of deformation in numerous deformation processes and in various metals. The most common processes that have been used are tension, compression, torsion, rolling and wire drawing. The stored energy after shock loading and cyclic loading has also been investigated.

Comparisons between the results of investigations in which different processes of deformation have been used present a problem because, in general, the states of strain differ from one deformation process to another. Although the stored energy is likely to be a function of all components of the state of strain as discussed in section 1.1, most values have been reported as a function of the largest normal or shear strain. Neither of these strains adequately defines the extent of deformation when different processes are to be compared.

An alternative parameter to the strain for defining the extent of deformation is the energy expended in the deformation process (see section 1.1). This parameter is always positive, which permits direct comparison of the processes of tension and compression; it also permits the comparison of torsion with these processes. On the other hand, in such processes as wire drawing and drilling, the expended energy has the disadvantage as a measure of the extent of deformation that it includes work done against friction. Also, the expended energy has not been reported for all investigations. In this review, strain and expended energy are used as alternative independent variables; preference is given to the latter when the effects of different processes of deformation are compared.

For a given metal and deformation process, the amount of stored energy increases with the extent of cold work. This increase was first reported by Hort[47, 48] in 1906 for iron, steel and copper deformed in tension. The first investigation in which the dependence of the stored energy on the extent of deformation was determined up to large strains was that of Sato[73] in 1931. His results for copper, deformed by torsion up to a strain $nd/l = 2.0$, are shown in Fig. 2 (p. 46) in the form of the stored energy E_s as a function of the expended energy E_w. Sato did not report the purity of the copper used.

Results for copper obtained by later investigators are included in Fig. 2. The large spread of values of the stored energy in this figure poses a problem. This spread increases as the extent of deformation increases: the ratio of the highest to the lowest value of the stored energy increases from approximately 2.9 at an expended energy of 50 cal/gram-atom to

TABLE 3.

Summary of Measured Values of

(See pages 44–45 for

Date	Investigators	Method[a]	Metal	Deformation process	Variables	Strain[b, c]
1900	Charbonnier and Galy-Aché[49]	1	Cu	Compression		
1906, 1907	Hort[47, 48]	1	Fe "tech. pure" Also steel Cu	Tension	$\bar{\varepsilon}$	0.034–0.109
c. 1922	Sinnatt[202]	1	Steel	Tension		
1925	Farren and Taylor[169]	1	Steel	Tension	$\bar{\varepsilon}$	0.05–0.12
			Cu	Tension	$\bar{\varepsilon}$	0.07–0.18
			Al	Tension	$\bar{\varepsilon}$	0.10–0.21
			Al	Tension	$\bar{\varepsilon}$	0.14–0.44
1925	Krivobok[93]	2-A	Steel	Compression	% carbon	c. 0.9
			Fe–1.68Si	Compression		
1928	Giraud[366]	1	Steel	Drawing	$\bar{\varepsilon}$	0.05–0.51
1929	Masima and Sachs[133]	1	Brass	Tension	$\bar{\varepsilon}$	0.01–0.04
1929	Smith[113]	2-R	Cu	Drawing		⩾3
			62.4Cu–37.6Zn	Drawing		⩾3
			Zn	Drawing		⩾3
1931	Sato[73]	2*	70Cu–30Zn	Torsion	nd/l	0.749
		2-A	Ag	Torsion	nd/l	1.23–2.19
			Cu	Torsion	nd/l	0.62–1.99
			Al	Torsion	nd/l	0.81–2.31
			Fe	Torsion	nd/l	0.83–1.14
			Ni	Torsion	nd/l	0.63–1.01
			Brasses	Torsion	nd/l, composition	0.21–1.84
			Bronzes	Torsion	nd/l, composition	0.23 0.68
			German silver	Torsion	nd/l	0.68, 1.31
1933	Rosenhain and Stott[41]	1	Al	Wire drawing		0.51
			Cu	Wire drawing		0.525
			Cu*	Wire drawing		0.25
1934	Taylor and Quinney[40]	(1) 1	Steel	Torsion	nd/l	0.046–0.366
			Decarb. steel	Torsion	nd/l	0.081–0.508
		(2) 1	Steel	Torsion	nd/l	0.067–0.70
			Decarb. steel	Torsion	nd/l	0.063–0.59
			Cu*	Torsion	nd/l	0.059–1.45
1934	France[114]	2-R	Fe	Rolling	$\bar{\varepsilon}$	
			Fe	Wire drawing		
			Cu	Wire drawing	$\bar{\varepsilon}$	
1934	Maier and Anderson[15]	1	Al	Wire drawing		
			Cu	Wire drawing		
1936	Kunin and Senilov[119]	1	Cu	Compression	$\bar{\varepsilon}$	0.08–0.50

THE STORED ENERGY OF COLD WORK

he Stored Energy of Cold Work
y to footnote symbols)

Strain rate	Grain size (mm)	$E_s^{(c,d)}$		$\frac{E_s}{E_w} \times 100^{(c,d)}$	Other measurements	Remarks
		(cal/g)	(cal/g-a)			
		≈ 0.9*	≈ 57*	13*		* Incidental to other work
		0.041–0.091	2.2–5.1	15–12		
				90		Doubtful validity
		0.054–0.156	3.0–8.7	c. 13*	Stress–strain	Each specimen
		0.022–0.073	1.4–4.6	10.5–8.0	curves	extended in several
		0.03–0.083	0.9–2.6	8–7		stages at a rate of
	Single cryst.	0·023–0.131	1.1–3.6	c. 5*		about 2 sec/stage
		> 0	> 0			* Approx. constant
	Single cryst.	> 0	> 0			
		1–9	60–500		Stress–strain curves*	Doubtful validity *Not reported
	Single cryst.			15–5		
		10.5	670			Doubtful validity
		9.0	580			
		6.7	440			
		1.08	69.1	12.6	Stress–strain curves	*Measured in calorimeter, and used to calibrate differential heating curves obtained from all other specimens
		0.23–0.31	25–33	3.9–2.6	curves in torsion,	
		0.18–0.32	11–20	4.3–1.8	anisothermal	
		1.79–3.60	48–97	34–17	kinetics	
		0.09–0.12	5.0–6.7	0.9–0.7		
		0.27–2.16	17–138	24–3†		† Function of composition and strain
		0.52 1.74	≈ 30–100	23–7†		
		2.68, 3.64	≈ 160, 230	22, 15		
3.53 m/min		0.11	3.0	1.2	U.T.S. of annealed and drawn wires	* Previously drawn to $\bar{\varepsilon} = 0.525$
4.37 m/min		0.23	15	3.1		
		≈ 0.03	≈ 2	0.5		
Total time 45 sec		0.04–0.76	2.2–42.4	13–9	Stress–strain curves	* Believed to contain 0.6% impurities
45 sec		0.05–0.67	2.8–37.4	13–7		
45 sec		0.161–1.267	9.0–70.7	11.6–8.7		
45 sec		0.047–0.66	26.2–36.9	8.1–7.7		
6–7 sec		0.019–1.149	1.21–73.0	9.6–8.2		
		2.1–4.8	120–270			
		7.3	410			
		≈ 0	≈ 0			
0.42 m/min		0.15	4.1	2 ±	Density	
0.42 m/min		0.23	15	9 ±		
0.07 mm/sec		0.04–0.29	2.6–18.7	15.0–8.3	Stress–strain curves	

(continued overleaf)

TABLE 3.1—continued

Date	Investigators	Method[a]	Metal	Deformation process	Variables	Strain[b,c]
1937	Quinney and Taylor[77]	(1) 2-A	Cu	Torsion		
			70Cu–30Zn	Torsion		
			Al	Torsion		
		(2) 2-A	Cu	Torsion		
			70Cu–30Zn	Torsion		
			Al	Torsion		
			Fe	Torsion		
			Ni	Torsion		
1941	Fedorov[120]	1	Cu	Compression	$\bar{\varepsilon}$	0.07–0.525
			Al	Compression	$\bar{\varepsilon}$	0.18–0.75
			Pb	Compression	$\bar{\varepsilon}$	0.25–0.69
			Sn	Compression	$\bar{\varepsilon}$	0.27–0.67
1946	Tizhnova[121]	1	Cu	Compression	$\bar{\varepsilon}$	0.20–0.51
			Cu–Ni alloys to 70% Ni	Compression	$\bar{\varepsilon}$, composition	0.10–0.69
1949	Epifanov and Rebinder[42]	1	Al	Drilling	Cutting fluids	
1949	Suzuki[14]	2-A	Cu 99.96	Compression	$\bar{\varepsilon}$	0.053–0.859
1950	Studenok[122]	1	Cu	Compression	Prior $\bar{\varepsilon}$ (compression) and $d\bar{\varepsilon}/dt$, $\bar{\varepsilon}$	0.025–0.48
1950	Degtiarev[123]	1	Cu	Compression	Prior $\bar{\varepsilon}$ (tension), $\bar{\varepsilon}$	0.09–1.02
1951	Kanzaki[76]	2-A	Cu 99.8	Compression	$\bar{\varepsilon}$	0.77, 1.05
			Ag 99.9	Compression	$\bar{\varepsilon}$	0.52, 0.79
			Al 99.9	Compression	$\bar{\varepsilon}$	0.92, 1.20
1951	Kanzaki[79]	2-A	Cu	Compression	γ, T	0.1–0.7*
1951	Bockstiegel and Lücke[367]	2-A	Cu	Wire drawing Torsion	$\bar{\varepsilon}$	5.6, 8.1 To fracture
1951	Bever and Ticknor[132]	2-R	75Au–25Ag	Rolling		c. 1.9
1952	Welber[78]	2-A	Cu	Torsion	nd/l	
1952	Borelius[100]	2-I	Al	Rolling		
			Cu	Rolling		
1952	Borelius et al.[101]	2-I	Al	Rolling		≈0.7
			Cu	Rolling		≈0.7
			Zn	Rolling		≈0.7
1952	Clarebrough et al.[75]	2-A	Cu 99.967	Torsion	nd/l	0.47–3.03
			Cu 99.988	Torsion	nd/l	0.47–2.25
1953	Bever and Ticknor[126]	2-R	75Au–25Ag	Rolling	$\bar{\varepsilon}$	0.17–2.53

THE STORED ENERGY OF COLD WORK 27

Strain rate	Grain size (mm)	$E_s^{(c,d)}$ (cal/g)	$E_s^{(c,d)}$ (cal/g-a)	$\frac{E_s}{E_w} \times 100^{(c,d)}$	Other measurements	Remarks
		0.32	20	7.0	Stress–strain curves in torsion*	* Not reported
		2.20	140	15.8		
		1.08	29	9.8		
		0.5	32	9.0		
		0.485	31.0	14.7		
		†	†			† Unreliable results
		1.2	67	15.0		
		0.78	45.8	5.6		
"Constant"		0.034–0.286	2.2–18.1	15–6	Stress–strain curves*	* Not reported
"Constant"		0.08–0.41	2.2–11.1	7–2		
"Constant"		≈ 0.0–0.004	≈ 0–0.8	2–1		
"Constant"		≈ 0.0–0.006	≈ 0–0.7	< 1		
"Constant"		0.13–0.48	8.3–30.5	18–8	Stress–strain curves*	* Not reported
"Constant"		0.18–1.06	11.3–63.7	32–9		
				3 ±		
	0.1	0.03–0.80	2–51		Elec. resistivity, hardness, microstructure, stress–strain curves, anisothermal kinetics	
0.05 mm/sec		0.0–0.246	0.0–15.7	0–15*	Stress–strain curves	Energy stored only after appreciable secondary deformation * Values of E_s/E_w do not correspond to those of E_s
"Constant"		0.0–0.67	0.0–42	0.0–13.8*	Load-compression curves	* See under 'Remarks' for Studenok
		0.34, 0.47	22, 30	5.4, 5.1	Anisothermal kinetics, electrical resistivity*, hardness*	* Not reported
		0.06, 0.09	6, 10	3.3, 2.3		
		0.84, 0.95	23, 26	11.0, 9.1		
	Single cryst.	0.006–0.082	0.4–5.2		Anisothermal kinetics	* Shear strain
		2.5, 2.7	159, 171			
		1.0	64			
		0.19–0.34	31–56			Preliminary to ref. 126
0.5 r.p.m.		0.23–0.41	15.9–28.4	2.8–3.4	Stress–strain curves*	* Not reported
						Preliminary to ref. 101 Investigation directed primarily at isothermal kinetics of energy release at 100°C
		0.100–0.331	6.4–21.0	2.6–1.0	Microstructure, X-rays, anisothermal kinetics	
		0.109–0.276	6.9–17.5	3.4–1.1		
		0.02–0.16	3–27	5–1.3	Stress–strain curves	

(continued overleaf)

TABLE 3.1—*continued*

Date	Investigators	Method[a]	Metal	Deformation process	Variables	Strain[b, c]
1953	Bever et al.[127]	2-R	75Au–25Ag	Orthogonal cutting	γ	1.12–3.78*
1953, 1954	Eugène[368, 369]	1	Steel Cu	Compression by impact		
1953	Clarebrough et al.[189]	2-A	Ni	Torsion		2.34
1953	Welber and Webeler[80]	2-A	Cu	Fatigue in tension*		
1954	Khotkevich et al.[87]	2*	Cd	Compression at 78°K	$\bar{\varepsilon}$	
			Pb	Compression at 78°K	$\bar{\varepsilon}$	
			Cu	Compression at R.T.	$\bar{\varepsilon}$	
1955	Clarebrough et al.[160]	2-A	Cu	Fatigue*		
1955	Clarebrough et al.[46]	2-A	Cu 99.96	Torsion	nd/l	?–1.87
			Cu 99.98	Compression	$\bar{\varepsilon}$	0.43–1.70
			Cu 99.55	Torsion	nd/l	0.47–2.53
			Cu 99.55	Tension		0.34
			Ni 99.6	Torsion	nd/l	0.94–2.34
1955	Leach et al.[134]	2-R	75Au–25Ag	Drilling at R.T. Drilling at 78°K		
1955	Åström[105, 106]	2-I	Al ≈ 99.99	Compression	$\bar{\varepsilon}$	0.09–0.60
1955	Gordon[104]	2-I	Cu 99.999+	Tension	$\bar{\varepsilon}$	0.10–0.33
1955	Bohnenkamp et al.[173]	2-A	Cu 99.96	Wire drawing	$\bar{\varepsilon}$	0.98–5.13
			Cu 99.88	Wire drawing	$\bar{\varepsilon}$	6.3
1956	Riggs[142]	2-A	Cu ≈ 99.99	Tension at R.T. and 78°K	T	0.20
				Torsion at R.T. and 78°K	T	0.26
1956	Greenfield and Bever[135]	2-R	82.6Au–17.4Ag	Drilling at R.T. Drilling at 78°K		

THE STORED ENERGY OF COLD WORK

Strain rate	Grain size (mm)	$E_s^{(c,d)}$ (cal/g)	$E_s^{(c,d)}$ (cal/g-a)	$\dfrac{E_s}{E_w} \times 100^{(c,d)}$	Other measurements	Remarks
Cutting speed 12 in./min		0.24–0.34	39–56	5.6–0.73		* Shear strain
		0.011	61	< 1		
		0.010	64	< 1	Hardness, electrical resistivity, density	More fully reported in ref. 46
	0.06–0.45	0.36–0.51†	23–32†			* Stress range 1000–20,000 lb/in.², 2000 cycles/min for 4 × 10⁴ cycles † Negative energy values
		1.13†	127	90–17	Electrical resistivity, anisothermal kinetics	* Pulse annealing † Values of E_s are upper limits
		0.53†	110	45–23		
		0.77†	49	17–3		
		0.04, 0.05	2.5, 3.2		Hardness, anisothermal kinetics	* 2000 cycles/min in push-pull; 0.25 × 10⁶ and 6.0 × 10⁶ cycles, respectively
		0.09–0.32	5.7–20.3	2.3–0.9	Electrical resistivity, density, X-ray diffraction, hardness, anisothermal kinetics	
		0.07–0.17	4.4–10.8	2.2–1.2		
		0.13–0.43	8.2–27.3	3.8–1.4		
		0.11	7.0	5.5		
		0.38–0.70	22.2–41.0	2.3–1.5		
		0.50*	82*	0.9	Stress–strain curves at R.T. and 78°K	* Values for straight-fluted drill
		1.48*	242*	1.4		
		0.12–0.27	3.16–7.28		Hardness, microstructure, isothermal kinetics	Rate equations fitted to the results
% elong./min	0.015	0.040–0.095	2.52–6.21	12.6–5.37	Hardness, microstructure, X-ray of grain size, isothermal kinetics	Rate equations fitted to the results
		≈ 0.3	≈ 19	2.5–0.4	Stress–strain curves, electrical resistivity, X-rays	
		1.5	95	1.6		
	0.02–0.05	0.072, 0.10	4.6, 6.4*	6.9, 7.5	Stress–strain curves, microhardness, X-ray diffraction	* First value applies to R.T.
	0.02–0.05	0.083, 0.16	5.3, 9.9*	4.5, 6.1		
		0.52	89	≈ 1	Stress–strain curves at R.T. and 78°K, X-ray diffraction	Release of stored energy during annealing also investigated
		1.19	206	≈ 1		

(*continued overleaf*)

TABLE 3.1—*continued*

Date	Investigators	Method[a]	Metal	Deformation process	Variables	Strain[b,c]
1956	Averbach et al.[148]	2-R	75Au–25Ag	Filing		
1956	Michell[188]	2-A	Ni	Grinding		
1956	Henderson and Koehler[88]	2-A	Cu 99.999 95.5Cu–4.5Zn	Compression at c. 90°K	$\bar{\varepsilon}$	0.81–1.24
1957	Michell and Haig[149]	2-A	Ni	Grinding		
1957	Greenfield and Bever[136]	2-R	Au–Ag alloys	Drilling at R.T.	Composition	
				Drilling at 78°K	Composition	
1957	Panin and Milevskaya[70]	1	Cu 99.90	Compression	$\bar{\varepsilon}$	0.16–0.36
			Ni 98.7	Compression	$\bar{\varepsilon}$	0.16–0.36
			Cu–Ni alloys	Compression	$\bar{\varepsilon}$, composition	0.16–0.36
			61.2Cu–38.61Zn	Compression		0.16–0.36
1957	Clarebrough et al.[161]	2-A	Cu, OFHC	Fatigue (bending and push–pull)	$\bar{\varepsilon}$, deformation process	
				Compression		0.69
				Fatigue following compression		
			Ni 99.6	Fatigue (push–pull)	No. of cycles	
1957	Clarebrough et al.[174]	2-A	Cu 99.98	Compression	$\bar{\varepsilon}$	0.36–1.21
1958	Wang and Brown[143]	2-A	Fe	Torsion	nd/l, T	1½–4½ turns
1958	Clarebrough et al.[124]	2-A	Cu 99.98	Compression	$\bar{\varepsilon}$, grain size	0.105–1.21 0.105–1.21
1959	Shermergor[72]	1	Steel	Compression	$\bar{\varepsilon}$, composition	0.02–0.30
1959	Titchener and Bever[128]	2-R	82.6Au–17.4Ag	Wire drawing at R.T.	$\bar{\varepsilon}$, grain size	0.36–4.81
				Wire drawing at R.T.	$\dot{\varepsilon}$	0.58
				Wire drawing at R.T.	$\dot{\varepsilon}$	3.46
			75Au–25Ag	Wire drawing at R.T.	$\bar{\varepsilon}$	0.58–3.65
			82.6Au–17.4Ag	Wire drawing at 78°K	$\bar{\varepsilon}$	0.36–2.05
				Wire drawing at 78°K	$\dot{\varepsilon}$	0.58

THE STORED ENERGY OF COLD WORK

Strain rate	Grain size (mm)	$E_s^{(c,d)}$ (cal/g)	$E_s^{(c,d)}$ (cal/g-a)	$\dfrac{E_s}{E_w} \times 100^{(c,d)}$	Other measurements	Remarks
		0.47	77		Hardness, X-ray diffraction	Release of stored energy during annealing investigated and correlated with structural changes Preliminary to Ref. 149
	0.1	0.13–0.19* 0.33* 3.6	8.3–12.1* 21.1* 210		Anisothermal kinetics X-ray diffraction, anisothermal kinetics	* Released up to c. 30°C
		0.11–0.68 0.51–1.48 0.22–0.58 0.25–0.56 0.45–0.86* 0.34–1.0 0.12, 0.03	22–94* 101–242* 14–37 15–33 28–53* 22–64 7.6, 1.9	0.35–1.1 0.92–1.36 16–12 19–13 34–22*	Stress–strain curves at R.T. and 78°K Stress–strain curves Hardness, X-ray diffraction, anisothermal kinetics	* Depending on composition * Values for 50Cu–50Ni $E_p > E_r$. No recrystallization in low-stress specimen
		0.13	8.3			Work softening in fatigue following compression
		0.05	2.9			E_s not a function of no. of cycles between 5×10^5 and 10^6. No recrystallization peak
		0.095–0.148	6.0–9.4		Density, electrical resistivity	
		0.15–1.1	8.4–61	4–8*	Heat capacity	* Values of E_s/E_w do not correspond to those of E_s
	0.030 0.150	0.04–0.19 0.02–0.19	2.5–12.1 1.3–12.1	10.7–1.8 6.5–1.8	Hardness, stress–strain curves, microstructure, anisothermal kinetics, recrystallization behavior	
.05/min in./min	0.018, 0.75	0.025–0.53 0.07–0.17*	1.4–30 11.9–28.8*	27–6* 4.10–0.29	Hardness	* dE_s/dE_w * No effect of grain size detected
03–1296 in./min	0.018	0.11–0.16†	19–27†		Hardness	† Passed through a maximum as function of $\dot{\varepsilon}$
17–1296 in./min	0.018	0.16–0.18†	28–31†			
n./min		0.16–0.25	26.6–40.2	3.24–0.54	Hardness	
⅟ in./min	0.018	0.33–0.93	57–160			
6 in./min	0.018	0.45, 0.36‡	78, 62‡			‡ 62 cal/g-atom at 6 in./min

(continued overleaf)

TABLE 3.1—*continued*

Date	Investigators	Method[a]	Metal	Deformation process	Variables	Strain[b,c]
1959	Kunin, V. N.[65]	1	Cu	Tension	$\bar{\varepsilon}$	0.02–0.32
			Ag	Tension	$\bar{\varepsilon}$	0.02–0.30
			Cd	Tension	$\bar{\varepsilon}$	0.02–0.20
1959	Schottky and Bever[177]	2-R	Ag 99.9+	Wire drawing		2.05
1959	Mima and Tokizawa[81]	2-A	Al 99.5	Compression	$\bar{\varepsilon}$	*0.29–1.51*
				Compression by multiple impact	$\bar{\varepsilon}$	*0.15–1.10*
1959	Kunin, V. N.[66]	1	Ag	Tension	$\bar{\varepsilon}$	0.02-fracture
			Cu	Tension	$\bar{\varepsilon}$	0.02-fracture
1959	Popov et al.[82]	2-A	97Al–3Cu	Compression by impact		*0.36*
			95Al–5Cu			*0.36*
1959	Iyer and Gordon[157]	2-I	Cu 99.999+	Compression following prior tension ($\bar{\varepsilon}_1 = 0.27$)	$\bar{\varepsilon}_2$	*$\bar{\varepsilon}_2 = 0.03$–$0.22$*
1959	Sizmann and Wenzl[94]	2-A	Cu electrolytic	Wire drawing		2.2
1960	Cohen and Bever[130]	2-R	Cu$_3$Au, initially disordered	Rolling	$\bar{\varepsilon}$	0.3–0.95
			Cu$_3$Au, initially ordered	Rolling	$\bar{\varepsilon}$	0.1–0.95
			Cu$_3$Au, initially disordered	Wire drawing	$\bar{\varepsilon}$	0.11–2.28
			Cu$_3$Au, initially ordered	Wire drawing	$\bar{\varepsilon}$	0.11–2.28
1960	Bailey and Hirsch,[107] Bailey[264]	2-I	Ag 99.99	Tension	$\bar{\varepsilon}$	*0.11–0.36*
1960	Michell and Lovegrove[171]	2-A	Ni 99.85	Compression in 3 mutually perp. directions		0.29 in each direction
1960	Titchener and Bever[159]	2-R	82.6Au–17.4Ag	Wire drawing at 78°K followed by wire drawing at R.T.	$\bar{\varepsilon}$	0.62 0.62–1.08

Strain rate	Grain size (mm)	$E_s^{(c,d)}$		$\dfrac{E_s}{E_w} \times 100^{(c,d)}$	Other measurements	Remarks
		(cal/g)	(cal/g-a)			
1.9 in./min		0.02–0.35	1.3–22	56–19		
1.9 in./min		0.01–0.18	1.1–19	71–21		
1.9 in./min		0.01–0.08	1.1–9	62–24		
		0.09	10			
High		0.94–1.77	25–48		Anisothermal kinetics, microstructure	Less energy stored at a given strain by multiple impact than by static compression
		0.63–1.45	17–39			
		0.01–0.18* †	1.1–19* †		Thermoelectric force, resistivity	* Values not given as a function of strain † Values not given Thermoelectric power and resistivity linear functions of E_s in both metals
		* 1.3 ± 0.66†	* 36 ± 17†		Apparent heat capacity‡	* Not reported. † Average of 6 runs ‡ As a function of prior heat treatment, deformation
		0.094–0.10*	5.99–6.5*		Hardness	* E_s passed through a minimum (5.1 cal/g-atom) at $\bar{\varepsilon}_2 = 0.1$. Work softening observed
		1.4	89		Anisothermal kinetics	Primarily a description of the calorimeter
	0.012	0.31–1.65	30–160		Energy of ordering, hardness, tensile tests, X-ray diffraction, electrical resistivity	
	0.012	0.72–3.92	70–380			
0.25 in./min	0.012	0.15–2.27	15–220	6–0.7		
0.25 in./min	0.012	0.15–6.60	15–650	4–1.5		
0.01/min	0.012	0.025–0.067	2.70–7.27*	10.9–4.1†	Transmission electron microscopy, stress–strain curves	* 7.75 cal/g-atom at $\bar{\varepsilon} = 0.28$ † Reported in ref. 108. Estimates of dislocation densities and energy of a dislocation
	0.03–0.04	0.28	16.4		X-ray diffraction, anisothermal kinetics	
in./min		0.52, 0.30*	90, 51*		Hardness	* Decrease due to annealing at R.T. † Decrease due to drawing at R.T. Work softening observed
in./min		0.30, 0.19†	51, 33†			

(continued overleaf)

TABLE 3.1—*continued*

Date	Investigators	Method[a]	Metal	Deformation process	Variables	Strain[b, c]
1960	Clarebrough et al.[179]	2-A	69Cu–31Zn	Torsion Tension	nd/l $\bar{\varepsilon}$	0.47, 1.87 *0.1–0.47*
1960	Pervakov et al.[195]	2-A	Ag 99.99	Compression* at R.T. Compression* at 77°K	$\Delta d/d_0$* $\Delta d/d_0$*	to 0.7 to 0.7
1960	Clarebrough et al.[170]	2-A	Ni 99.96 Ni 99.85	Compression Compression Torsion	$\bar{\varepsilon}$	*0.1–1.21* 1.21 2.01
1960	Williams[38]	1	Cu	Compression by impact Compression by impact		0.2–0.67 0.2–0.67
1960	Hertsriken et al.[96]	2-A	Cu 99.98 Cu-0.74Zn Cu-0.92Zn Armco Fe	Torsion Torsion Torsion Torsion	nd/l nd/l nd/l nd/l	0.6, 1.3 0.6 1.3 0.8
1961	Appleton et al.[150]	2-R	82.6Au–17.4Ag	Shock loading	Shock pressure	0.08–0.26*
1961	Loretto and White[166]	2-A	Cu 99.999	Compression	Grain size	*0.36*
1961	Clarebrough et al.[175]	2-A	Al 99.991	Compression		*1.39*
1961	van den Beukel[89]	2-A	Cu 99.999 Ni 99.99	Compression at 77°K Compression at 77°K	$\bar{\varepsilon}$ $\bar{\varepsilon}$	*0.40–1.08* *0.60–0.87*
1961	Williams[51]	1	Cu 99.999 Cu-0.012Ag Cu-0.08Ag Cu-1.0Ag Cu 99.999	Compression by impact Compression by impact Compression by impact Compression by impact Compression by impact	$\bar{\varepsilon}$ Grain size	0.19–0.67 0.3 0.3 0.3 0.3
1961	Roessler and Bever[137]	2-R	Cu$_3$Au, initially disordered Cu$_3$Au, initially ordered	Wire drawing at 78°K Wire drawing at 78°K	$\bar{\varepsilon}$ $\bar{\varepsilon}$	0.18–2.4 0.37–2.4

THE STORED ENERGY OF COLD WORK 35

Strain rate	Grain size (mm)	$E_s^{(c,d)}$ (cal/g)	$E_s^{(c,d)}$ (cal/g-a)	$\overline{E_w} \times 100^{(c,d)}$	Other measurements	Remarks
		0.84, 2.59 −0.09*–0.81	53.8, 153 −5.8*–51.9		Resistivity, density, hardness, aniso-thermal kinetics	* Negative value. Includes energy absorbed in reordering but not energy released during recrystallization
		to 0.45	to 49	15–5	Electrical resistivity†	* Lateral compression of wires
		to 0.7	to 76	40–8		† $E_s/(\Delta\rho/\rho)$ independent of strain and temperature
		* 0.30 *	* 17.6 *		Density, electrical resistivity, aniso-thermal kinetics	* Values not stated. Estimates of dislocation densities
High		0.10–0.28*	6.3–18.0*	10–4‡	Release of energy immediately after impact, stress–strain curve	* Measured 0.01 min after deformation
High		0.09–0.23†	5.9–14.5†			† Measured 1.0 min after deformation
						‡ Time of measurement not stated. Some qualitative observations on a Cu–Zn and a Cu–Al alloy
		0.4, 1.0	25, 65	10, 8	Anisothermal kinetics	
		1.5	95	29		
		2.0	130	14		
		1.3	75	8		
Very high		0.03–0.18†	5–31†		Microhardness, microstructure	* Calculated transient strain
						† Shock pressures 120–510 kbar
	0.70	0.085	5.4		Hardness, aniso-thermal kinetics	
	0.15	0.116	7.4			
	1	0.12	3.2		Resistivity, hardness, anisothermal kinetics	
		0.11–0.28*	7.0–18*		Anisothermal kinetics	* Between 90° and 300°K
		0.07–0.12†	4.1–7.0†			† Between 90° and 220°K. Preliminary to ref. 91
$\times 10^4$/min	0.2	0.09–0.23*	6–14.5*	9.6–4.0	Stress–strain curve, kinetics of release immediately after deformation	* At 1 min after deformation
$\times 10^4$/min	0.1	0.162	10.3			
$\times 10^4$/min	0.08	0.168	10.7			
$\times 10^4$/min	0.05	0.181	11.5			
$\times 10^4$/min	0.01–0.4	0.162–0.137	10.3–8.7			
in./min	0.036	0.7–6.5	70–630		Resistivity, microhardness, microstructure, energy after R.T. annealing	
in./min	0.036	1.5–9.8	150–950			

(continued overleaf)

TABLE 3.1—continued

Date	Investigators	Method[a]	Metal	Deformation process	Variables	Strain[b,c]
1962	Clarebrough et al.[176]	2-A	Ag 99.98 Au 99.9	Compression Compression		1.39 1.39
1962	Bell and Krisement[109,110]	2-I	Ni 99.9	Compression at 213°K	$\bar{\varepsilon}$	*0.32, 0.64*
1962	Bailey and Hirsch[108]	2-I	Ag 99.99	Tension		*0.23*
1962	Williams[52]	1	Pb 99.999	Compression by impact	$\bar{\varepsilon}$	0.4–1.7
			Al 99.9	Compression by impact	$\bar{\varepsilon}$	0.35–1.3
			Al 99.99	Compression by impact	$\bar{\varepsilon}$	0.3–1.3
			Ag 99.99	Compression by impact	$\bar{\varepsilon}$	0.18–0.69
			Ni 99.38	Compression by impact	$\bar{\varepsilon}$	0.11–0.43
			Ni 99.9	Compression by impact	$\bar{\varepsilon}$	0.17–0.63
			Fe 99.95	Compression by impact	$\bar{\varepsilon}$	0.08–0.42
			Fe 99.96	Compression by impact	$\bar{\varepsilon}$	0.11–0.34
			Zr 99.9	Compression by impact	$\bar{\varepsilon}$	0.13–0.35
1962	Dillon[61]	1	Al, 1100 F	Cyclic torsion	Strain amplitude	0.001–0.006
1962	Dillon[62]	1	Al, 1100 O	Cyclic torsion	Strain amplitude, cycles	
1962	Iyer and Gordon[151]	2-I	Cu, tough-pitch Cu 99.999	Shock loading Shock loading		
1963	Gordon[370]	2-I	Cu 99.999	Tension		*0.28*
1963	White, J. L.[125]	2-A	Cu 99.999	Compression at 0°C	$\bar{\varepsilon}$, grain size	to 0.4
1963	Erdmann and Jahoda[143b]	1	65Ni–35Cu	Tension at 4.2°K	$\bar{\varepsilon}$	0.09–0.20
1963	Vandermeer and Gordon[172]	2-I	Al-0.016Cu	Rolling at 0°C		0.51
1963	van den Beukel[91]	2-A	Cu 99.999	Compression at 77°K	$\bar{\varepsilon}$	0.41–1.08
			99.9Cu–0.1Au	Compression at 77°K	$\bar{\varepsilon}$	*0.75–1.19*
			99.0Cu–1.0Au	Compression at 77°K	$\bar{\varepsilon}$	*0.72–0.99*

THE STORED ENERGY OF COLD WORK

Strain rate	Grain size (mm)	$E_s^{(c,d)}$ (cal/g)	$E_s^{(c,d)}$ (cal/g-a)	$\dfrac{E_s}{E_w} \times 100^{(c,d)}$	Other measurements	Remarks
	0.5	0.145	15.6		Resistivity, hardness, density, anisothermal kinetics	
	0.5	0.037	7.3			
	1.4	0.22, 0.36	13, 21	6.5, 4.2	Hardness, isothermal kinetics	E_p and E_r distinguished. Rate equations determined
	0.050	0.050	5.40	6.4	Electron microscopy, X-ray diffraction, isothermal kinetics	Primarily an investigation of recrystallization kinetics
High		0.007–0.016	1.4–3.3	2–1*	Stress–strain curves, X-ray diffraction	* $(dE_s/dE_w) \times 100$ † Different lots.
High	0.05	0.089–0.30	2.4–8.2	3–2*		Energy release measured between 0.01 and 10 min after deformation.
High	0.10	0.093–0.35	2.5–9.3	3–2*		Values of E_s in table are for 10 min except for 1 min for Pb.
High	0.01	0.084–0.23	9–25	8–2*		
High	0.05	0.12–0.39	7–23	9–2*		
High	0.05	0.12–0.26	7–15	7–1*		Kinetics of release in first 10 min analyzed
High	0.02	0.13–0.54†	7.5–30†	8–?*		
High	0.10	0.09–0.27†	5–15†	6–4*		
High	0.5	0.24–0.83	22–76	16–10*		
		0.01*	0.27*	6–28*		* Average over one cycle at a shear strain amplitude of 0.006
		0.003*	0.08*	15*		* Average over one cycle at a shear strain amplitude of 0.003
Very high		*	*		Hardness, isothermal kinetics, X-ray diffraction	* No absolute values of E_s; percent release reported as function of time
Very high		*	*			
		0.098	6.25		Isothermal kinetics	
	0.035, 0.40	0.12, 0.075*	7.5, 4.8*			* At $\bar{\varepsilon} = -0.4$. Value for fine grained material given first
		0.055–0.50*		49–38	Stress–strain curves	* Cal per specimen
	0.5	0.0098*	0.62*		Microstructure, isothermal kinetics	* Before recrystallization
		0.11–0.28*	7–18*	2.8–2.0	Specific heat, anisothermal kinetics	* Released up to 0°C † Released up to 0°C; considerable scatter
		0.18–0.36†	11–23†			
		0.24–0.40†	16–26†			

(continued overleaf)

TABLE 3.1—continued

Date	Investigators	Method[a]	Metal	Deformation process	Variables	Strain[b, c]
1963	van den Beukel[91] (cont.)	2-A	99.9Cu–0.1Ag	Compression at 77°K	$\bar{\varepsilon}$	0.66–0.95
			99.0Cu–1.0Ag	Compression at 77°K	$\bar{\varepsilon}$	0.59–0.89
			Ag 99.999	Compression at 77°K	$\bar{\varepsilon}$	0.58–0.99
			Au 99.999	Compression at 77°K	$\bar{\varepsilon}$	0.69–1.19
			Ni 99.99	Compression at 77°K	$\bar{\varepsilon}$	0.61–0.87
			Al 99.99	Compression at 77°K		
1963	Appleton and Bever[138]	2-R	82.6Au–17.4Ag	Torsion, at R.T.	nd/l	0.25–1.05
				Torsion at 78°K	nd/l	0.25–1.45
				Torsion at 4.2°K	Strain rate	0.60
				Torsion	T	0.54
1963	Wenzl[95]	2-A	Cu 99.999	Rolling	$\bar{\varepsilon}$	0.18–2.2
			Cu 99.98	Rolling	$\bar{\varepsilon}$	0.18–2.0
			99.7Cu–0.3As	Rolling	$\bar{\varepsilon}$	0.19–2.04
1963	Loretto et al.[90]	2-A	Cu 99.99	Wire drawing at R.T.		0.44
				Wire drawing at 77°K		0.44
1963	Åström[190]	2-I	Mo 99.98	Compression	$\bar{\varepsilon}$	0.11–0.51
1963	Williams[43]	1	Cu*	Tension	$\bar{\varepsilon}$	0.03–0.35
1963	White and Koyama[74]	2-A	Cu 99.999	Compression	$\bar{\varepsilon}$, grain size	0.248 0.352 0.392
			Cu, OFHC	Compression	Lot, grain size	0.40
1963	Dillon[63]	1	Al, 1100-O	Cyclic torsion	Strain amplitude	0.007, 0.01†
1963	Williams[53]	1	85.5Cu–14.5Al	Compression by impact	$\bar{\varepsilon}$	0.09–0.23
				Compression by impact	$\bar{\varepsilon}$	0.20–0.55
			70Cu–30Zn	Compression by impact	$\bar{\varepsilon}$, prior heat treatment	0.09–0.39
1963	Scattergood et al.[152]	2-R	82.6Au–17.4Ag	Shock loading Wire drawing	Shock pressure $\bar{\varepsilon}$	0.09–2.03

Strain rate	Grain size (mm)	$E_s^{(c,d)}$ (cal/g)	$E_s^{(c,d)}$ (cal/g-a)	$\frac{E_s}{E_w} \times 100^{(c,d)}$	Other measurements	Remarks
		0.14–0.24†	9–15†		Specific heat, aniso-thermal kinetics	† Released up to 0°C; considerable scatter
		0.16–0.23†	10–15†			* Released up to 0°C
		0.06–0.17*	6–18*			
		0.08–0.13*	16–26*			
		0.07–0.11*	4–6*			
		‡	‡			‡ Not reported
$\dot{\gamma}_{max} = 0.2$/min		0.06–0.12	10–21	12–5	Stress–strain curves, hardness	* Maxima in curves of E_s v$\bar{\gamma}_{max}$ and E_w v$\bar{\gamma}_{max}$
$\dot{\gamma}_{max} = 0.2$/min		0.15–0.81	25–140	16–8		
$\dot{\gamma}_{max} = 0.03$ to 10/min		*	*			† From 4.2°K to 100°C
$\dot{\gamma}_{max} = 0.2$/min		1.1–0.10†	190–18†		Hardness, micro-structure, aniso-thermal kinetics	* Rolling speed
1 cm/min*	0.017	0.165–0.331	10.5–21.0	2–0.2†		† Values do not correspond to values of strain
1 cm/min*	0.010	0.072–0.344	4.6–21.6	1.3–0.2†		
1 cm/min*	0.007	0.060–0.260	3.8–16.5	1.3–0.2†		
		0.145	9.2		Resistivity*, aniso-thermal kinetics from c. 90°K	* Of 99.99 and 99.999 Cu
		0.55	35			
		0.027*	2.6*		Isothermal kinetics, energy released after neutron irradiation	* Released in successive anneals at 80°, 98° and 186°C after $\bar{\varepsilon} = -0.51$ at R.T.
0.05 in./min		0.02–0.23	1–14.5	27–4†	Stress–strain curve	* "High purity" † $(dE_s/dE_w) \times 100$. Primarily a description of equipment
	0.01	0.077	4.9	5.6	Anisothermal kinetics	* Depending on lot. Primarily a description of the calorimeter
	0.02	0.11	6.7	4.7		
	0.30	0.066	4.2	3.4		
	0.02	0.087, 0.15*	5.5, 9.3*	3.7, 5.7*		
	0.12	0.068	4.3	2.6		
		0	0	0	Hysteresis loops	"Steady state" hysteresis loop † Shear strain amplitude
High		0.36–0.98*	19–52*		Flow stress	* Slowly cooled from 500°C
High		0.41–1.58†	22–84†	26–10§		† Quenched from 800°C
High	0.04‡	0.28–1.23‡	18–79‡	27–10§		‡ As received § $(dE_s/dE_w) \times 100$
Very high		0.02, 0.14*	3, 24*		Microhardness, microstructure, isochronal annealing kinetics	* At 70 and 270 kbar
		0.02–0.15	3–25			

(continued overleaf)

TABLE 3.1—continued

Date	Investigators	Method[a]	Metal	Deformation process	Variables	Strain[b,c]
1963	Hargreaves et al.[158]	2-A	Cu 99.99	Prior extension ($\bar{\varepsilon}_1 = 0.29$) followed by secondary compression	$\bar{\varepsilon}_2$	0.0 0.05 0.17
1964	Erdmann and Jahoda[143a]	1	Cu–Ni alloys	Tension at 4.2°K	Composition	0.01–0.095
1964	Williams[50]	1	Cu 99.999	Tension	$\bar{\varepsilon}$	0.05–0.59
1964	Beardmore et al.[153]	2-R	Cu$_3$Au, initially disordered	Shock loading	Shock pressure	0.11–0.22*
			Cu$_3$Au, initially ordered	Shock loading	Shock pressure	0.11–0.25*
1964	Kunin et al.[67]	1	Cu 98.3	Tension	$\bar{\varepsilon}$	0.002–0.02
1965	Erdmann and Jahoda[144]	1	98Cu–2Ni	Tension at 4.2°K	$\bar{\varepsilon}$	~0–0.003
			92Pb–8In*	Tension at 4.2°K	$\bar{\varepsilon}$	~0–0.01
1965	Chuang and Bever[187]	2-R	Au 99.999+	Wire drawing at 78°K		1.05
1965	Nakada[68]	1	Al	Compression	γ, orientation	0.03–0.65
			Ag	Compression	γ, orientation	0.06–0.36
			Al	Compression	$\bar{\varepsilon}$	0.01–0.18
1965	Williams[54]	1	Cu 99.999+	Tension	Grain size	0.01–0.35 0.06–0.28
1965	Robinson and Bever[140]	2-R	AgMg	Torsion at 25° and 165°C	nd/l, composition	0.01–0.10
1965	Bell[111]	2-I*	Ni 99.9	Compression at 213°K		0.63
1965	Taoka et al.[97]	2-A	Fe-2.74Si	Rolling	Orientation	1.2
			Fe-2.98Si	Rolling		1.9
			Fe-0.004C	Rolling		1.9
			Fe-0.003C	Rolling		1.9
1966	Dillon[64]	1	Cu, OFHC	Cyclic torsion	Cycles	0.005†
1966	Ebel and Lugscheider[99]	2-A	Al 99.99	Rolling		2.3
1966	Robinson and Bever[183]	2-R	TlBi$_2$	Torsion	nd/l	0.06, 0.31, 0.44

Strain rate	Grain size (mm)	$E_s^{(c,d)}$ (cal/g)	$E_s^{(c,d)}$ (cal/g-a)	$\dfrac{E_s}{E_w} \times 100^{(c,d)}$	Other measurements	Remarks
		0.141	9.0		Density, electron microscopy, aniso-thermal kinetics	E_s passed through a minimum of 5.9 cal/g-a at $\bar{\varepsilon}_2 = -0.05$
		0.092	5.9			
		0.145	9.2			
		*	*	40–60†	Thermal conductivity	* Not reported † Max. value at approx. 60% Cu
	Single crysts	*0.0001–0.046*	*0.007–2.9*		Stress–strain curves	
Very high		*1.2–1.8*	115–170		Resistivity, microhardness, microstructure	* Calculated max. transient strain
Very high		*1.4–4.0*	135–390			
0.07/min		*0.0005–0.006*	*0.032–0.38*	20–53*	Stress–strain curve	* Maximum value at $\bar{\varepsilon} = 0.005$
				73–32.5	Thermal conductivity	* Heavily cold-worked at room temperature
2.9 in./min		0.5	99		Resistivity, annealing kinetics	
High	Single crysts	*0.007–0.19**	*0.2–5.0**	40–2.4*	Stress–strain curves	* Primarily a function of γ; relatively insensitive to orientation except for Al$\langle 100 \rangle$
High	Single crysts	*0.0014–0.019**	*0.15–2.0**	40–20*		
High	0.2	*0.015–0.19*	*0.4–5.0*	40–19		
0.025/min	0.06, 0.13	*0.002–0.21**	*0.1–13**		Stress–strain curves, X-ray diffraction, release of energy immediately after deformation	* No effect in changing grain size from 0.06 to 0.13 mm † $(dE_s/dE_w) \times 100$
0.025/min	0.76	*0.023–0.16*	*1.8–10*	18–6†		
0.22/min	0.017	*0.13–7.2*	9–480	9.7–33*	Resistivity, hardness	* At 25°C
					Isothermal release kinetics	Concerned primarily with energy release * Some anisothermal annealing
	Single crysts	3.9, 7.3*	210, 394*		Resistivity, X-ray diffraction, hardness, anisothermal kinetics, flow stress	* Depending on orientation
		4.5	244			
		1.8	101			
		1.5	♂			
		*0.003**	3*	20*	Stress–strain curves in torsion, hysteresis loops	* Average over one cycle and seven specimens † Shear strain amplitude
					X-ray diffraction, anisothermal release kinetics	Concerned primarily with relation between stored energy and lattice strain
0.22/min	0.05	*0.024–0.125*	5–26		Resistivity, hardness, stress–strain curves	

(*continued overleaf*)

TABLE 3.1—*continued*

Date	Investigators	Method[a]	Metal	Deformation process	Variables	Strain[b, c]
1966	Halford[69]	1	Cu, OFHC, annealed	Cyclic torsion	γ range, cycles	0.006–0.08
		1	Cu, OFHC, cold-worked	Cyclic torsion	γ range, cycles	0.008–0.08
		2-A	Cu, OFHC, annealed	Cyclic torsion	γ range, cycles	0.006§, 0.02‖
		2-A	Cu, OFHC, cold-worked	Cyclic torsion	γ range, cycles	0.008¶, 0.024** 0.026‡
1967	Panin *et al.*[71]	1	Cu	Tension	$\bar{\varepsilon}$	0.14–0.26
			98Cu–2Al	Tension	$\bar{\varepsilon}$	0.14–0.26
			95.5Cu–4.5Al	Tension	$\bar{\varepsilon}$	0.14–0.26
			92.5Cu–7.5Al	Tension	$\bar{\varepsilon}$	0.14–0.26
			91Cu–9Ni	Tension	$\bar{\varepsilon}$	0.14–0.34
			90Cu–10Zn	Tension	$\bar{\varepsilon}$	0.14–0.34
			64Cu–36Ni	Tension	$\bar{\varepsilon}$	0.14–0.26
			61Cu–39Zn	Tension	$\bar{\varepsilon}$	0.14–0.34
1967	Ham[39]	1	Cu 99.99	Bending		0.14*
			Cu 3.8Sn	Bending		0.14*
1967	Wolfenden[55, 56]	1	Cu 99.999	Tension	γ, orientation	0.005–0.5
1967	Kovacs[83]	2-A	Cu	Torsion	Shear strain*	0.6, 1
			Al	Torsion	Shear strain*	0.65
			Ag	Torsion	Shear strain*	—
1967	Brillhart *et al.*[154]	2-I	Cu 99.9	Shock loading	Shock pressure	0.22, 0.26*
1967	Chin and Grant[84, 85]	2-A	Cu 0.3Al$_2$O$_3$*	Extrusion at 760°C		3.3†
			Cu 1.1Al$_2$O$_3$*			3.3†
1967	Waldman and Bever[181]	2-R	Ag–Cd alloys	Wire drawing	$\bar{\varepsilon}$, composition	0.5–2.6
1968	Lugscheider and Wildhack[92]	2-A	Al 99.99	Tension at 78°K	$\bar{\varepsilon}$	0.095 to ~0.40
1968	Smith and Bever[129]	2-R	Au 99.99	Wire drawing	$\bar{\varepsilon}$, T	0.5–3.2* 0.5–1.75†
1968	Wolfenden and Appleton[146, 147]	1	Cu 99.97	Tension at 78°K	$\bar{\varepsilon}$, orientation	0.1–0.375
			Al 99.99	Tension at 78°K	$\bar{\varepsilon}$, orientation	0.1–0.45
			Cu 99.97	Tension at 78°K	$\bar{\varepsilon}$	0.1–0.3
			Al 99.99	Tension at 78°K	$\bar{\varepsilon}$	0.1–0.5
1968	Erdmann and Jahoda[145]	1	70Cu–30Ni	Tension at 4.2° to 70°K	T	
1969	Wolfenden[57]	1	Ag 99.92	Tension	γ, orientation	0.05–0.37
1969	Wolfenden[59]	1	82.6Au–17.4Ag	Tension	$\bar{\varepsilon}$, grain size	0.008–0.09
1969	Gangulee and Bever[180]	2-R	97Ag–3Mg	Wire drawing	$\bar{\varepsilon}$	0.35–1.0
			94.6Ag–5.4Mg	Wire drawing	$\bar{\varepsilon}$	0.35–1.0
			93Ag–7Mg	Wire drawing	$\bar{\varepsilon}$	0.35–1.45
			93Ag–7Mg*	Wire drawing	$\bar{\varepsilon}$	0.35–1.45

Strain rate	Grain size (mm)	$E_s^{(c,d)}$ (cal/g)	(cal/g-a)	$\dfrac{E_s}{E_w} \times 100^{(c,d)}$	Other measurements	Remarks
	5	~ 0–0.04*	~ 0–2.5*	15–2.5*	Shear stress amplitude, stress–strain curves in tension, hardness	* Depending on strain range and cycles
		0 to −0.05†	0 to −3.1†	5 to −12.5†		† Released energy
						§ 500. ‖ 89 cycles
	5	0.07§, 0.12‖	4.4§, 7.6‖			¶ 1000. ** 260 cycles
		0.03¶, 0.05**	1.9¶, 3.2**			‡ 1000 cycles
		0.08‡	5.1‡			
		0.13–0.27	8–17		Stress–strain curves	
		0.23–0.48	14–30			
		0.45–1.05	27–63			
		0.43–1.3	25–75			
		0.16–0.43	10–27			
		0.25–0.6	16–38			
		0.49–1.07	30–66			
		0.47–1.56	30–100			
3000/min				19	Kinetics of immediate energy release	* A mean strain
3000/min				24		
0.01/min	Single crysts	0–0.064	0–4.0	12–27	Shear stress	
		0.24, 0.4	15.2, 25.4		Anisothermal kinetics	* Not well defined
		0.34	9.2			
		0.48	51.6			
	0.013	0.21, 0.24	13.2, 15.0		Resistivity, density, X-ray diffraction, electron microscopy	* Calculated max. transient strain
55 in./min		0.42	26.2		Anisothermal kinetics	* By volume
55 in./min		0.41	25.6			† Extrusion ratio of 28:1
3 in./min		0.42–1.9	50–210		Resistivity	
	0.06	0.04–0.2*	1.2–6.0*	8.2–4.8*	Anisothermal kinetics	* Only the energy released up to room temperature was measured
		0.08–0.1*	16–20*		Resistivity, yield stress in tension	* Room temperature
		0.3–0.86†	60–170†			† 78°K
0.03/min	Single crysts	0.05–0.5	3.0–30	~ 100–20	Tensile stress	
0.03/min	Single crysts	0.11–1.4	3.0–37.5	~ 100–15		
0.03/min	0.05	0.09–0.2	6.0–12.5	~ 100–13		
0.03/min	0.1	0.3–0.7	8.0–19	40–23		
				68–0		Deformed in successive steps at increasing temperatures
0.04/min	Single crysts	0–0.018	0–2.0	2.3–13.6	Shear stress	
0.05/min	0.003, 0.045	0–0.016	0–2.63		Stress–strain curves	
		0.38–0.82	37–80		Resistivity	* Initially long-range ordered
		0.71–1.54	65–140			
		0.86–2.12	75–185			
		1.84–4.43	160–385			

(*continued overleaf*)

TABLE 3.1—continued

Date	Investigators	Method[a]	Metal	Deformation process	Variables	Strain[b, c]
1969	Khotkevich and Sirenko[178]	2*	Cd 99.95+	Compression at 77°K	$\bar{\varepsilon}$	0.2–0.9
			80Cd–20Pb	Compression at 77°K	$\bar{\varepsilon}$	0.2–0.9
			60Cd–40Pb	Compression at 77°K	$\bar{\varepsilon}$	0.2–0.9
			60Pb–40Cd	Compression at 77°K	$\bar{\varepsilon}$	0.2–0.9
			80Pb–40Cd	Compression at 77°K	$\bar{\varepsilon}$	0.2–0.9
			Pb 99.95+	Compression at 77°K	$\bar{\varepsilon}$	0.2–0.9
1969	Wolfenden[60]	1	Al 99.996	Tension	γ, orientation	0.028–0.117
1969	Bogachev and Denisova[139]	2*	Cu	Torsion at 77°K, 203°K and 20°C	nd/l, T	0.25–0.75
1971	Wolfenden[182]	1	Cu$_3$Au	Tension	$\bar{\varepsilon}$, order	0.04–0.35
1971	Wolfenden[131]	1	Cu 99.999	Tension	$\bar{\varepsilon}$	0.025–0.20
1971	Filatovs and Schwaneke[371] Filatovs and Leighly[372]	2-A	Cu 99.999+	Tension	$\bar{\varepsilon}$	0.11–0.34

[a] Method: 1 denotes single-step methods; 2 denotes two-step methods; A denotes anisothermal annealing; I denotes isothermal annealing; R denotes reaction methods.

[b] Strain: For torsion figures represent values of nd/l. For all other cases they represent true strain unless otherwise indicated. Note, however, that the negative sign has been omitted from compressive strains.

approximately 5.6 at an expended energy of 800 cal/gram-atom. An appreciable part of this spread probably arises from differences in the concentration and identity of impurities, as discussed in subsection 3.2.2. Smaller contributions to the spread may arise from differences in grain structure (subsection 3.2.1) and rate of deformation (subsection 3.1.3). It appears unlikely, however, that at a given strain the combined effects of all known variables can cause the highest and lowest values of the stored energy to differ by more than a factor of about 3.

Beyond an expended energy of about 250 cal/gram-atom the curves of stored energy vs. expended energy tend to fall into two distinct groups representing relatively high and relatively low values of the stored energy; only the curve of Khotkevich et al.[87] lies between the two groups. The curves in the lower group level off progressively as the extent of deformation increases, while those in the upper group show no or only a slight tendency to do so.

The high values include nine sets obtained by single-step methods[40, 43, 65, 70, 119–123] and three obtained by annealing methods.[14, 76, 96] Not all of these values are plotted in Fig. 2. Of the three sets of data obtained by annealing, those of Suzuki[14] and Kanzaki[76] must be treated with reserve for the reasons that will be given in subsection 4.1.1. Among the low values only those of Williams[38, 51, 54] (not plotted in Fig. 2) were obtained by a single-step method of measurement, and they are among the highest in the low group.

Strain rate	Grain size (mm)	E_s [c,d]		$\frac{E_s}{E_w} \times 100$ [c,d]	Other measurements	Remarks
		(cal/g)	(cal/g-a)			
		0.2–1.2	22.5–135	90–15		* Pulse annealing
		0.16–0.8	21.0–105			
		0.11–0.6	16.5–90	80–27		
		0.09–0.47	15.2–79.3			
		0.07–0.35	13.2–65.9			
		0.03–0.25	6.2–51.7	80–15		
0.05/min	Single crysts	0.0004–0.026*	0.01–0.7*			* Error cited as between 100 and 58%
		0.1–0.7† 0.4–1.35‡ 1.6–2.8§	6.35–44.5† 25.4–86.0‡ 101–177§		Recrystallization kinetics	* Pulse annealing † 20°C ‡ 203°K § 77°K
0.02/min				0–25	Stress–strain curves	
	0.18	*0–0.1*	*0–40*			
	0.06–0.5	*0.004–0.16*	*0.24–10.0*		Stress–strain curves	
		0.030–0.083	*1.9–5.3*			

[c] Strain, E_s, E_s/E_w: Figures in italics are converted for this review; all others are values reported by authors.
[d] E_s, E_s/E_w: Where range is shown, the limits correspond to the limits shown under "Strain", unless otherwise noted.

Thus the single-step methods generally appear to yield values of the stored energy that are higher than those obtained by annealing methods. Probable reasons for this are given in section 2.1.

In the group of low values are those of Clarebrough et al.,[46,75,124] J. L. White and Koyama[74] and White[125] (not plotted in Fig. 2) obtained by anisothermal annealing experiments, and those of Gordon[104] obtained by isothermal annealing experiments. The agreement among the results of these investigators is satisfactory; any differences can be accounted for by the effects of differences in grain size as discussed in subsection 3.2.1 and differences in the concentration and identity of impurities as discussed in subsection 3.2.2. The progressive leveling with increasing extent of deformation as shown by the curves in the lower group was also found by Bever and co-workers using a reaction method with gold–silver alloys,[126–128] gold[129] and the alloy Cu_3Au.[130]

Williams[38,43,51–54] showed that an appreciable amount of energy was released immediately after deformation. This release has since also been observed by Ham.[39] With 99.999% copper the amount released between 0.01 and 10 min after the end of the deformation ranged from approximately 5% of the stored energy at low strains to approximately 25% at a compressive strain $\bar{\varepsilon} = -0.66$.[51] The value of the energy 10 min after the end of deformation was still 40% above the value obtained by Gordon[104] with 99.999% copper at an expended energy of 115 cal/gram-atom and 80% above the value

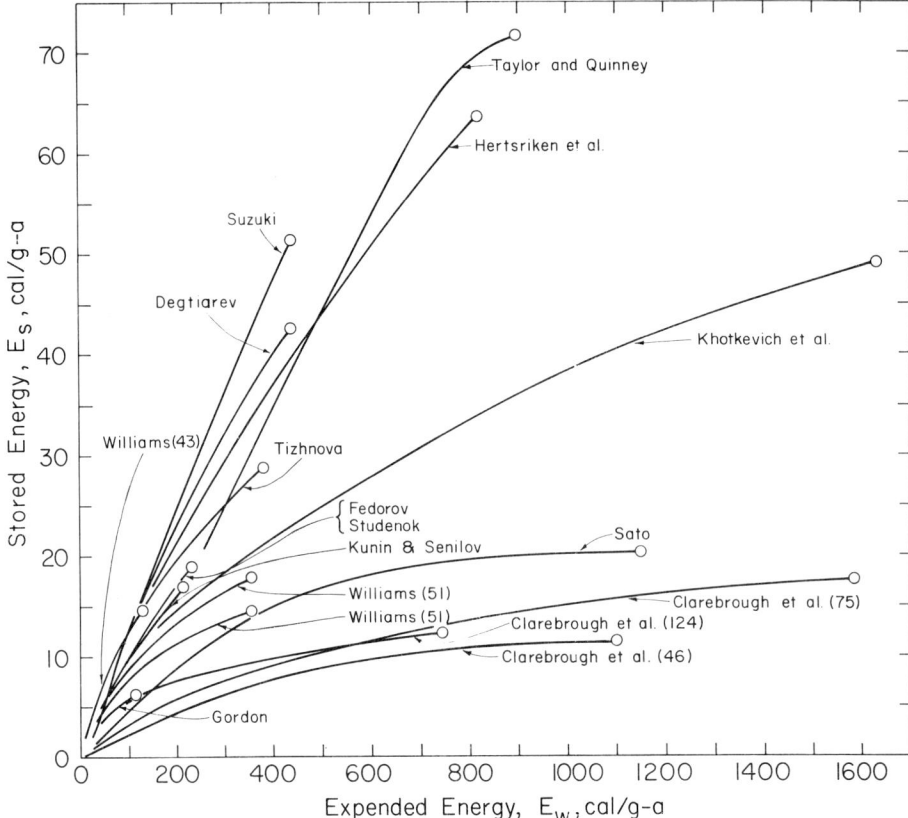

Fig. 2. Stored energy as a function of expended energy in copper, reported by various investigators. The point at the end of each curve represents the last experimental point obtained in each investigation. Clarebrough et al.,[46, 75, 124] Degtiarev,[123] Fedorov,[120] Gordon,[104] Hertsriken et al.,[96] Khotkevich et al.,[87] Kunin and Senilov,[119] Sato,[73] Suzuki,[14] Studenok,[122] Taylor and Quinney,[40] Tizhnova,[121] Williams[51] at 0.01 (upper curve) and 1 min (lower curve), Williams.[43]

obtained by J. L. White and Koyama[74] with 99.999% copper at an expended energy of 149 cal/gram-atom. Williams[54] suggested that the difference may have arisen because in other investigations the energy released by recovery during annealing was not measured. Wolfenden concurred with this interpretation and supported it with additional data for copper[131] which he considered to be equivalent to those of Williams. The possibility cannot be dismissed, however, that a small systematic error was present in the values of the expended energy E_w or the heat evolved Q, from which Williams' values of the stored energy were derived. It should be mentioned that Wolfenden used a different apparatus from that of Williams.

It was shown in section 2.1 that, in single-step methods, small errors in the quantities E_w and Q can lead to large percentage errors in the values of the stored energy. It was also pointed out that systematic errors in the one quantity do not cancel systematic errors in the other except fortuitously, and that the errors most likely to occur in both quantities lead to inflated values of the stored energy. A change of 2% in the quantities E_w and Q measured by Williams would reduce his values of the stored energy to agree with those of

Gordon. The presence or absence of systematic errors in Williams' values, however, cannot be determined from the available evidence.

The release of energy immediately after deformation as observed by Williams[38, 43, 51–54] appears to be insufficient to lower the curves of the high group to agree with those of the low group in Fig. 2. Differences in the identity, concentration and distribution of impurities may account for an appreciable part of the difference in the stored energies. It is possible that the copper used by Taylor and Quinney and the Russian investigators contained specific impurities that accounted for part of the differences between their values and those in the low group. The balance of these differences is likely to be due to systematic errors in the measurement of the quantities E_w and Q. A total systematic error of 6% would account for the difference between the high curves and the curve of Clarebrough et al.[124] for 99.98% copper. However, the possibility that the high values are correct cannot be dismissed. Support for them is provided by the values of Hertsriken et al.,[96] which were obtained by anisothermal annealing.

In this review the low values of the stored energy shown in Fig. 2 are accepted as being substantially correct; this statement applies to values which include energy released immediately after deformation, such as the values reported by Williams.[51–54] The high values are rejected as probably incorrect or at least not generally representative of the stored energy in copper. In some instances, however, the high values can be used to consider the effect of specific variables, such as strain rate, on the stored energy.

These considerations apply also to the values of the energy stored by metals other than copper. The dependence of the stored energy on the extent of deformation for other metals is similar to that for copper and may be seen for nickel, iron, silver and aluminum in Figs. 3 to 6, which will be discussed in subsection 3.2.3.

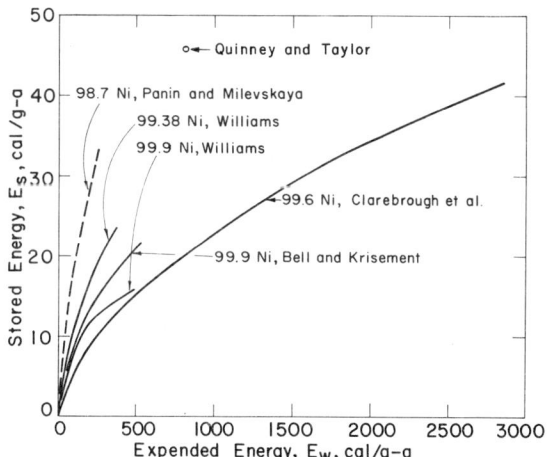

FIG. 3. Stored energy as a function of expended energy in nickel, reported by various investigators. Bell and Krisement,[109] Clarebrough et al.,[46] Panin and Milevskaya,[70] Quinney and Taylor,[77] Williams.[52]

Curves of the stored energy as a function of the energy expended in deforming an alloy are shown in Fig. 7. These curves were obtained by Bever and co-workers for a 75Au–25Ag alloy deformed by rolling,[126, 132] orthogonal cutting[127] and wire drawing.[128] Their shape resembles that of the lower curves for copper.

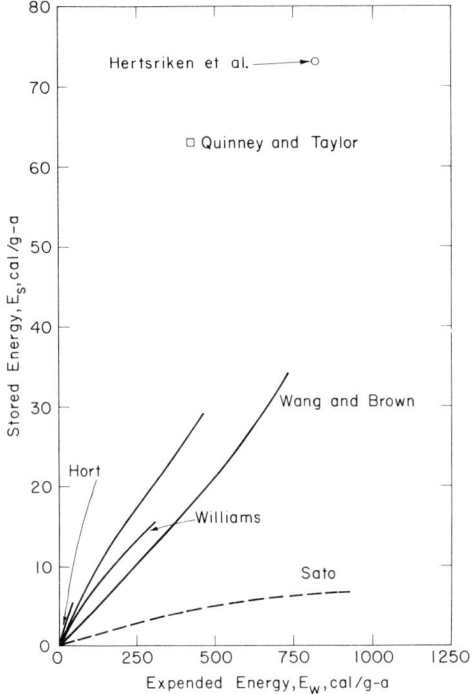

FIG. 4. Stored energy as a function of expended energy in iron, reported by various investigators. Hertsriken et al.,[96] Hort,[47,48] Quinney and Taylor,[77] Sato,[73] Wang and Brown,[143] Williams.[52]

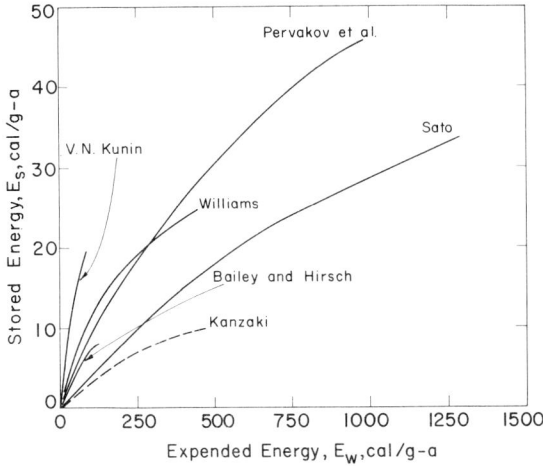

FIG. 5. Stored energy as a function of expended energy in silver, reported by various investigators. Bailey and Hirsch,[108] Kanzaki,[76] V. N. Kunin,[65] Pervakov et al.,[195] Sato,[73] Williams.[52]

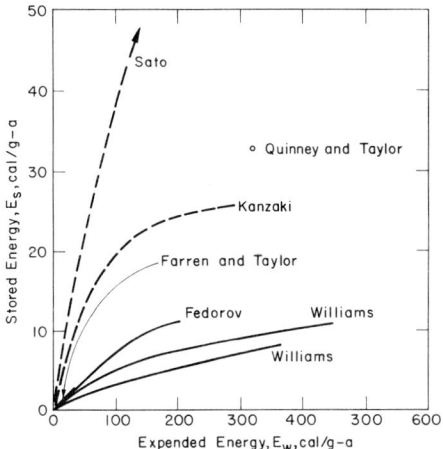

FIG. 6. Stored energy as a function of expended energy in aluminum, reported by various investigators. Farren and Taylor,[169] Fedorov,[120] Kanzaki,[76] Quinney and Taylor,[77] Sato,[73] Williams.[52]

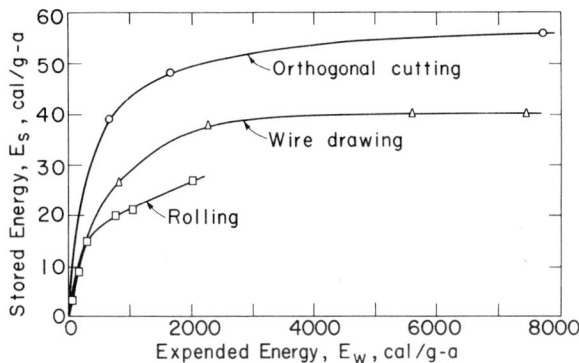

FIG. 7. Stored energy as a function of expended energy in a 75Au–25Ag alloy during different processes of deformation. Orthogonal cutting, Bever, Marshall and Ticknor;[127] wire drawing, Titchener and Bever;[128] rolling, Bever and Ticknor.[126, 132]

The use of strain, whether normal or shear, instead of expended energy as a measure of the extent of deformation does not alter the general shape of the curves. This is illustrated by Fig. 8 in which some of the data of Fig. 2 are plotted against the nondimensional quantity, nd/l, which is $1/\pi$ times the maximum shear strain in torsion.

An exception to the rule that the stored energy increases with increasing strain has been reported by Masima and Sachs,[133] who observed periodic decreases in the energy during the extension of single crystals of brass. These decreases occurred whenever there was a change from one operative slip system to another. Confirmation of these findings would be desirable.

3.1.1.1. *The Ratio of the Stored to the Expended Energy*

The amount of the stored energy may be considered in relation to the amount of energy expended during deformation. The relation between the two quantities can be expressed as the ratio E_s/E_w, or as the instantaneous rate of increase of the stored energy with respect

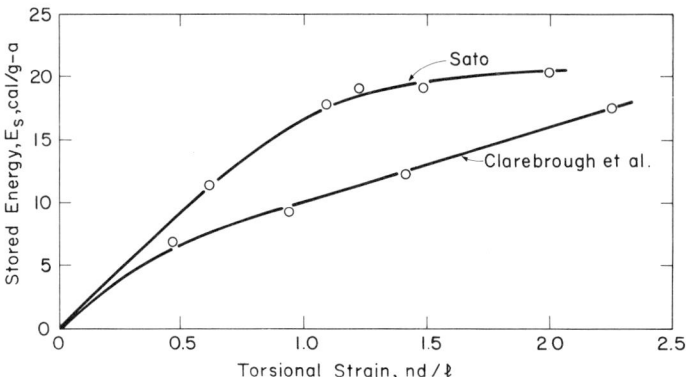

Fig. 8. Stored energy as a function of torsional strain in copper. Sato,[73] Clarebrough et al.[75]

to the expended energy dE_s/dE_w. The latter will be called in this review the instantaneous rate of energy storage. Williams[51] has pointed out that

$$\frac{dE_s}{dE_w} = \frac{1}{V_m \bar{\sigma}} \frac{dE_s}{d\bar{\varepsilon}} \qquad (3.1)$$

where V_m is the gram-atomic volume, $\bar{\sigma}$ the flow stress in appropriate units and $dE_s/d\bar{\varepsilon}$ the gradient of the curve of stored energy plotted against true strain. The relation in this form holds for deformation by uniaxial loading.

In single-step methods of measuring the stored energy, the deformation is usually carried out in successive stages, and the ratio of the incremental amounts of energy stored δE_s and expended δE_w in a stage has approximately the same value as the instantaneous rate of storage dE_s/dE_w at the mean strain in that stage. The ratio $\delta E_s/\delta E_w$ will be called in this review the incremental rate of energy storage.

The ratio E_s/E_w was reported by Charbonnier and Galy-Aché (1900),[49] the first investigators of the stored energy (see Table 3.1) and has continued to receive attention to the present time. The ratio is a small quantity, seldom as high as 15% and usually less than 10% for deformation at room temperature. It depends on the extent of the deformation. It is largest at low strains and decreases in a generally hyperbolic manner with increasing strain. Typical curves are shown in Fig. 9 for two gold–silver alloys deformed by wire drawing at room temperature. Similar curves have been obtained with other metals and other methods of deformation, as may be seen in Fig. 10.[46] The values of the ratio E_s/E_w for gold–silver alloys, copper and nickel are roughly equal at the same values of the expended energy, although the frictional losses occurring in wire drawing inflate the values of E_w for the gold–silver alloys. One investigation[67] found a maximum in E_s/E_w and another[53] in $\delta E_s/\delta E_w$ at small strains. Further work on these ratios at small strains is desirable.

Plotting the instantaneous rate of energy storage dE_s/dE_w or the incremental rate of storage $\delta E_s/\delta E_w$ against the extent of deformation provides sensitive tests for any tendency of the stored energy to attain a limiting value. Taylor and Quinney[40] used the incremental rate in an attempt to demonstrate that copper deformed by torsion at room temperature became "saturated" with stored energy at a value of the expended energy approximately

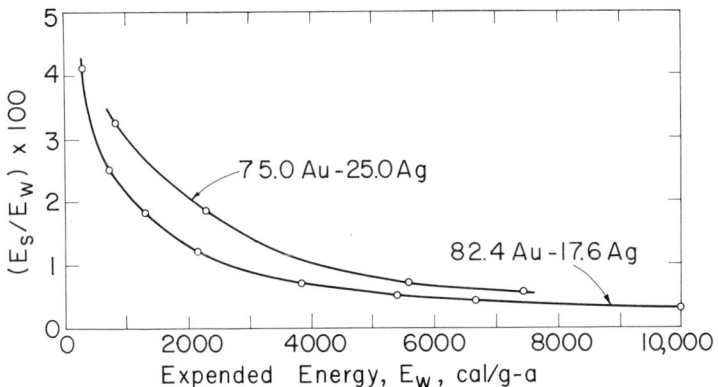

FIG. 9. The ratio $(E_s/E_w) \times 100$ as a function of expended energy in two gold–silver alloys. Titchener and Bever.[128]

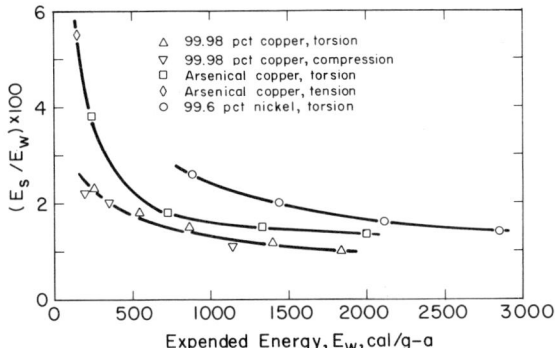

FIG. 10. The ratio $(E_s/E_w) \times 100$ as a function of expended energy in copper and nickel. Clarebrough et al.[46]

equal to that at which the indentation hardness reached a limiting value in specimens deformed by compression.

Williams, in a series of investigations, presented curves showing the incremental rate $\delta E_s/\delta E_w$ as a function of strain for copper[43, 51] and for lead, aluminum, silver, nickel, iron and zirconium.[52] His curves for copper and silver are shown as typical examples in Fig. 11. The rate of storage in all metals decreased progressively with increasing strain. Williams' results suggest that, while the ability of these metals to store additional energy decreases as the strain increases, a true saturation level of the stored energy may not exist for them.

3.1.2. *Temperature of Deformation*

Kanzaki[79] was the first to carry out stored energy measurements after deformation below room temperature. He compressed single crystals of copper at 93° and 193°K. His energy measurements began only at room temperature, however, and thus could not establish the full effect of a low temperature of deformation.

The first systematic investigation of the effect of temperature of deformation on the

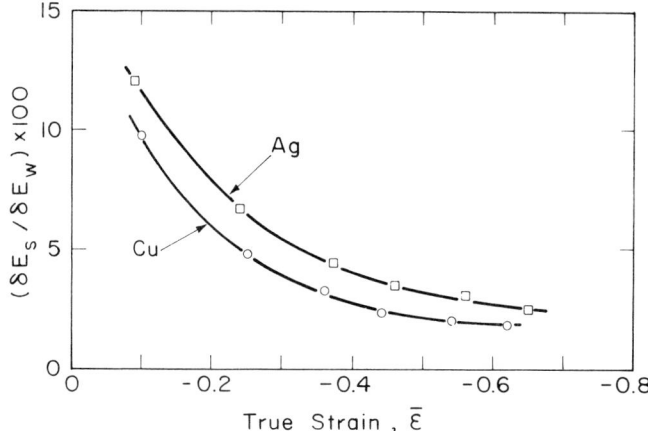

FIG. 11. The incremental rate of energy storage as a function of true strain in copper and silver deformed by compression. Williams.[38, 52]

stored energy was made by Leach, Loewen and Bever[134] with chips prepared by drilling a 75Au–25Ag alloy at 78°K† and at room temperature. The energy stored at 78°K was 242 cal/gram-atom compared with 82 cal/gram-atom at room temperature. An investigation by Greenfield and Bever[135] with an 82.6Au–17.4Ag alloy yielded values of 206 and 89 cal/gram-atom for chips prepared by drilling at 78°K and room temperature. The effect of the temperature of cold work on the energy stored in drillings of gold–silver alloys containing up to 50% (65 at. %) silver is shown in Fig. 12.[136]

In an investigation of the stored energy in which drilling was employed, the extent of deformation could not be a controlled variable. The energy stored as a function of strain in wires of an 82.6Au–17.4Ag alloy drawn at 78°K and at room temperature was investigated by Titchener and Bever.[128] The results are shown in Fig. 13. In agreement with the results of Greenfield and Bever,[135, 136] the energy stored at 78°K was several times that stored at room temperature. The rate of energy storage decreased at large strains at 78°K, but to a much smaller degree than at room temperature. In an investigation of 99.99% gold wires drawn at 78°K, Smith and Bever[129] found that the stored energy was a linear function of strain up to the largest strain attained, namely $\bar{\varepsilon} = 1.73$, whereas at room temperature the stored energy reached a nearly constant lower value at high strains. Roessler and Bever[137] found that the energy stored by initially ordered and initially disordered alloy Cu_3Au deformed at 78°K was larger than the energy stored during deformation at room temperature.[130] These results will be discussed in 3.2.4.1.

The dependence of the stored energy on the temperature of deformation was investigated over the range 4.2°K to 100°C by Appleton and Bever,[138] who deformed an 82.6Au–17.4Ag alloy by torsion to a strain $nd/l = 0.54$. The results are shown in Fig. 14. The experimental difficulties in making energy measurements at 4.2°K were considerable, so that some uncertainty attends the values for this temperature, as noted by the authors.

† In this review, the figure of 78°K for the boiling point of liquid nitrogen as originally reported by Bever and co-workers has been retained in discussing their investigations; this figure is also retained if it was used by other investigators.

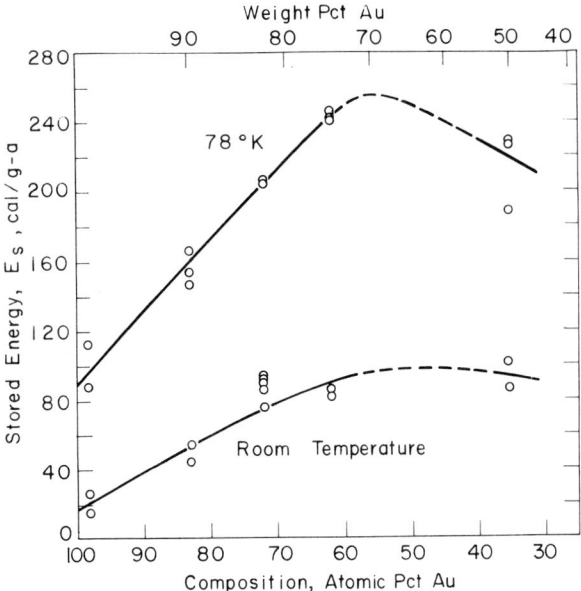

Fig. 12. The stored energy as a function of composition in gold–silver alloys deformed at 78°K and room temperature. Greenfield and Bever.[136]

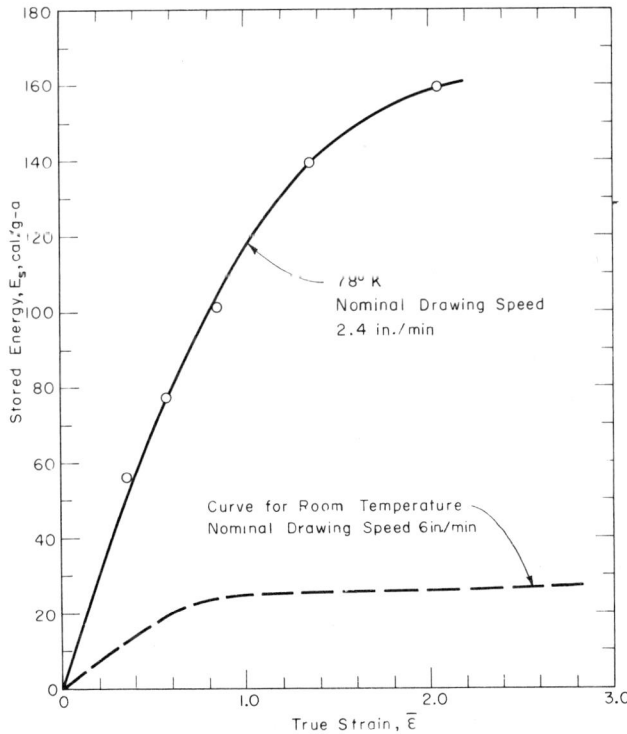

Fig. 13. The stored energy as a function of true strain in wires of an 82.6Au–17.4Ag alloy drawn at 78°K and room temperature. Titchener and Bever.[128]

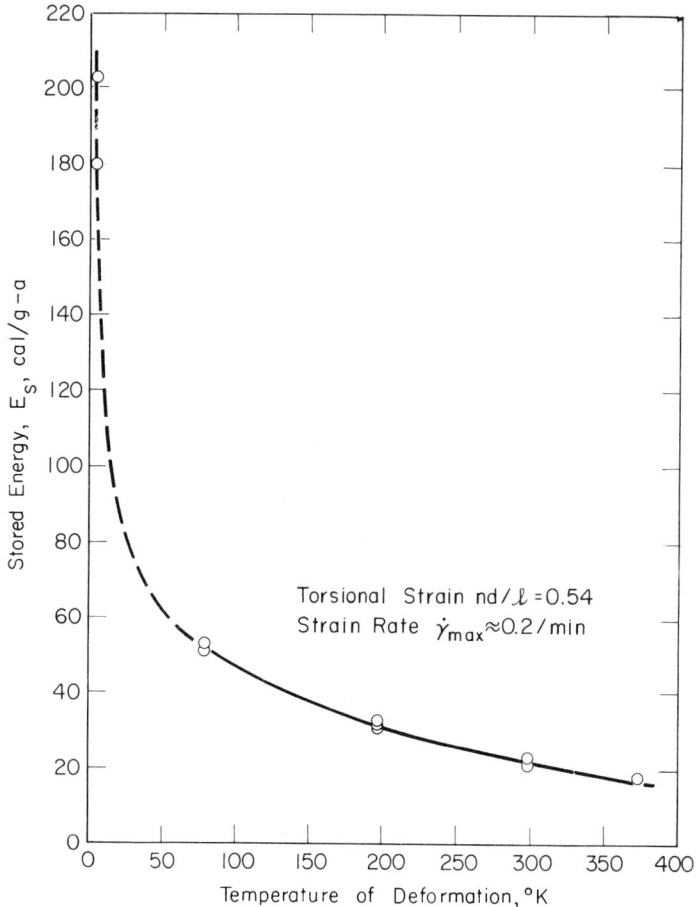

Fig. 14. The stored energy as a function of temperature of deformation in an 82.6Au–17.4Ag alloy deformed by torsion. Appleton and Bever.[138]

Their values of the stored energy increased sharply as the temperature of deformation was reduced from 78° to 4.2°K. The large magnitude of the energy stored at 4.2°K compensated to some extent for the larger errors.

Khotkevich et al.[87] measured the energy stored in cadmium and lead deformed at 77°K, but not at room temperature. Their values of the stored energy must be derived from their reported values of the ratio of the stored to the expended energy. The value for lead (34 cal/gram-atom) may be compared with the values for lead heavily deformed at room temperature reported by Fedorov[120] (≈ 0.8 cal/gram-atom) and Williams[52] (≈ 3 cal/gram-atom). Loretto et al.[90] measured the energy stored in 99.99% copper wires drawn to a reduction of area of 54% ($\bar{\varepsilon} = 0.78$) at the temperature of liquid nitrogen and at 20°C. The values of the stored energy were 35 and 9.2 cal/gram-atom. Bogachev and Denisova[139] measured the stored energy in copper deformed at 77° and 203°K and 20°C. At a torsional strain $nd/l = 0.75$, the stored energies were 177, 86, and 44.5 cal/gram-atom, respectively. Curves of the stored energy versus strain had less tendency to level off at the lower temperatures. These investigations show that the stored energy in nominally pure metals, as in gold–silver alloys and in Cu_3Au, increases with decreasing temperature.

Robinson and Bever[140] found that the energy stored in the compound AgMg increased linearly with strain when the deformation was at 25°C, but reached a plateau in deformation at 165°C. Their results are discussed further in 3.2.4.1.

In several investigations of metals deformed below room temperature, only that part of the stored energy which was released up to room temperature or somewhat above was measured. Henderson and Koehler[88] in an investigation of 99.999% copper compressed at about 90°K, carried their energy measurements to about 40°C; similar measurements on alpha brass were carried to about 80°C. Van den Beukel's measurements of the energy stored in aluminum, silver, copper, gold, nickel, and dilute alloys of silver in copper and gold in copper were made from 98°K to about room temperature.[89, 91] Lugscheider and Wildhack[92] measured the energy released by aluminum after deformation at 78°K; they carried these measurements up to room temperature. The investigations of Henderson and Koehler, van den Beukel, and Lugscheider and Wildhack, which throw much light on low-temperature annealing kinetics, will be discussed in Part 4.

Riggs[142] deformed copper and Wang and Brown[143] Armco iron at 78°K and room temperature. Like Kanzaki, however, these investigators began their measurements only at room temperature. Their investigations, therefore, gave limited information on the effect of the temperature of deformation on the stored energy.

The ratio E_s/E_w increases with decreasing temperature of deformation. This may be seen for an 82.6Au–17.4Ag alloy in Fig. 15.[138] The results of Khotkevich et al.[87] for copper

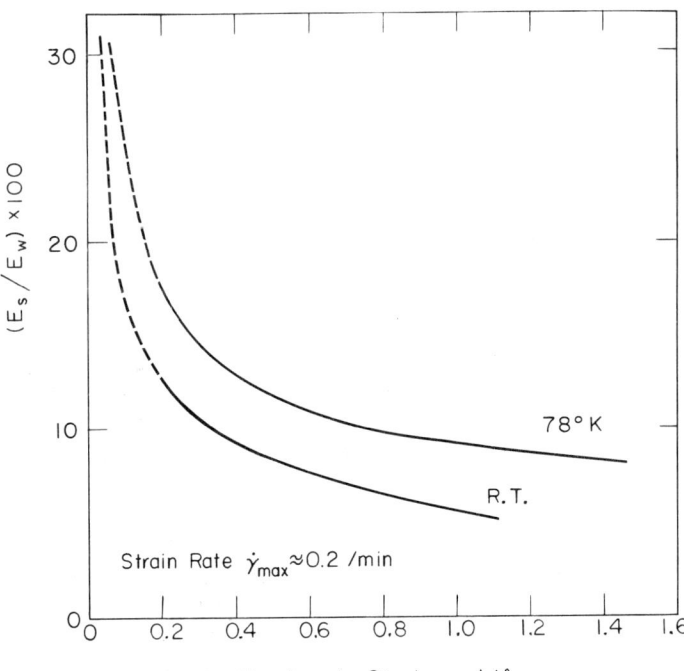

FIG. 15. The ratio $(E_s/E_w) \times 100$ as a function of strain in an 82.6Au–17.4Ag alloy deformed by torsion at 78°K and room temperature. Appleton and Bever.[138]

deformed at room temperature and cadmium and lead deformed at 78°K are also consistent with an increase in the ratio E_s/E_w with decreasing temperature. In an investigation of the energy stored in copper–nickel alloys during tensile straining at 4.2°K, Erdmann and Jahoda[143a] observed relatively high values of E_s/E_w; these values were in the range 40–60%. In another investigation, they found values of E_s/E_w as high at 73% at small strains in a 98Cu–2Ni alloy.[144]

Erdmann and Jahoda[145] reported the incremental rate of energy storage $\delta E_s/\delta E_w$ for a 70Cu–30Ni alloy at temperatures between 4.2° and 70°K. The results are shown in Fig. 16. The rate did not decrease monotonically as the temperature increased. Since the specimen was deformed in steps at successively higher temperatures, the results may reflect the release of some part of the energy by a work-softening mechanism.

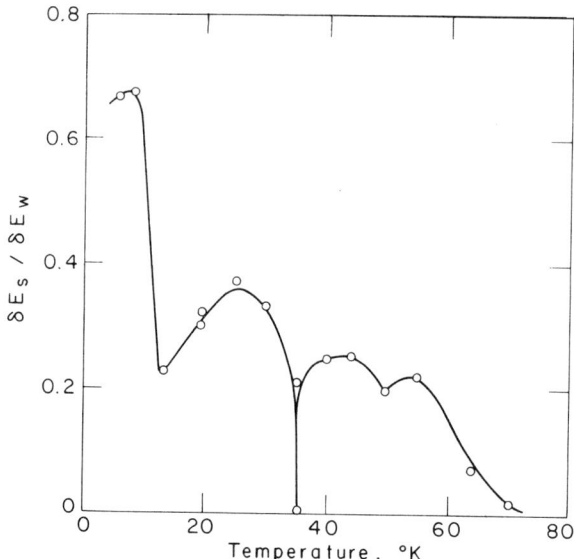

FIG. 16. The incremental rate of energy storage $\delta E_s/\delta E_w$ in a 70Cu–30Ni alloy deformed in tension at successively higher temperatures. Erdmann and Jahoda.[145]

Wolfenden and Appleton[146] and Wolfenden[147] deformed single crystals of copper and aluminum and polycrystalline copper in tension at 78°K. At the smallest strains at which they measured the stored energy, namely $\bar{\varepsilon} = 0.1$, they reported values of E_s/E_w of approximately 100%. As discussed in subsection 3.2.1, such high values of E_s/E_w are probably incorrect.

3.1.3. *Rate of Deformation*

The effect of the rate of deformation was investigated by Titchener and Bever[128] on wires of an 82.6Au–17.4Ag alloy drawn at room temperature. The results are shown in Fig. 17. The stored energy did not change with strain rate up to drawing speeds of 6 in./min. Above that speed, it increased sharply to a maximum at about 216 in./min, beyond which it fell rapidly. Neither the wire nor the dies were cooled. The decrease in stored energy at drawing speeds above 216 in./min was believed to be the result of partial recovery promoted

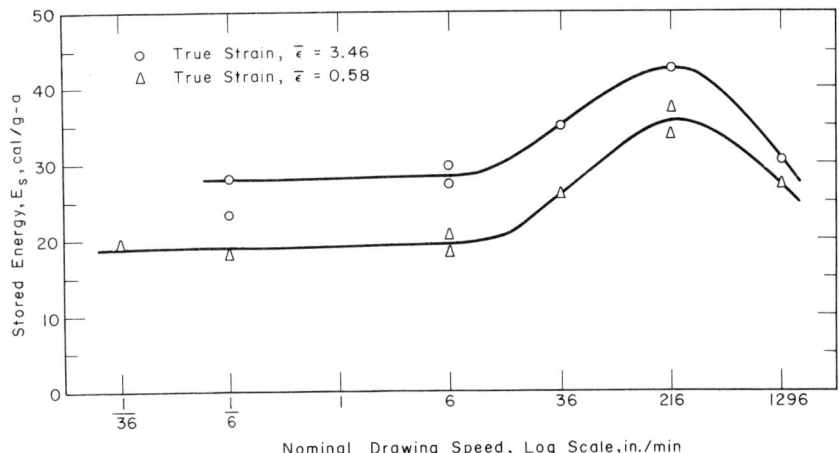

FIG. 17. The stored energy as a function of drawing speed in wires of an 82.6Au–17.4Ag alloy drawn at room temperature. Titchener and Bever.[128]

by heating at high drawing speeds. Wires drawn at 78°K at a speed of 6 in./min stored one-quarter to one-third less energy than wires drawn to the same strain at a speed of 2.4 in./min.

The energy E'_s retained at 78°K after torsion at 4.2°K was measured by Appleton and Bever.[138] This energy, shown in Fig. 18 as a function of the maximum shear strain rate $\dot{\gamma}_{max}$, passed through a maximum at $\dot{\gamma}_{max} \approx 0.3$. The energies stored in deformation at 78°K and room temperature, however, were independent of the maximum shear strain rate in the neighborhood of $\dot{\gamma}_{max} = 0.2$ per min.

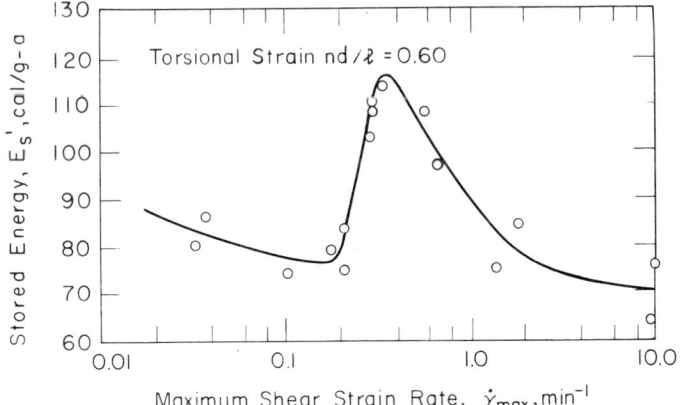

FIG. 18. The stored energy measured at 78°K as a function of maximum shear strain rate in an 82.6Au–17.4Ag alloy deformed by torsion at 4.2°K. Appleton and Bever.[138]

The literature appears to contain no other reports of extensive investigations aimed primarily at determining the effect of strain rate on the stored energy. The rates of deformation in most investigations may be assumed to have been low. In some cold-working processes, however, the rates of shear are high. Values of the stored energy obtained from drillings, filings and grindings have invariably been much larger than those obtained from metals deformed in processes having slower rates of deformation. For example, Greenfield and Bever[136] found a stored energy of 85 cal/gram-atom in chips prepared by drilling a 75Au–25Ag alloy at room temperature. In the same laboratory Averbach et al.[148] measured a value of 76 cal/gram-atom in filings produced from the same alloy at room temperature. These values may be compared with the largest values of 27 cal/gram-atom stored in the same alloy after rolling[126] and 40 cal/gram-atom after wire drawing.[128] Michell and Haig[149] with 99.6% nickel obtained a value of 210 cal/gram-atom for the energy stored in grindings compared with the highest value of 41 cal/gram-atom measured by Clarebrough et al.[46] in specimens of the same batch deformed by torsion. The large values of the energy stored in drillings, filings and grindings may not be entirely or even primarily due to the high strain rate characteristic of these processes. The ratios of surface to volume are much larger than those in wire or strip, and the higher values of the energy may be partly due to faster cooling after deformation and the resulting reduction in recovery effects. Also the paths of the deformation differ as discussed in section 1.1.

Williams[54] compared values obtained by him for the stored energy in copper deformed in tension at a strain rate of 0.22 min^{-1} with previous results for copper compressed in his impact calorimeter at an estimated strain rate of 40,000 min^{-1}. The stored energy at the slow rate was found to be slightly higher than that at the high rate. Williams concluded that the effect of strain rate was small. The uncertainties in comparing absolute values of the stored energy obtained by different methods have already been mentioned (subsection 3.1.1).

The effects of shock loading will be discussed in subsection 3.1.4.

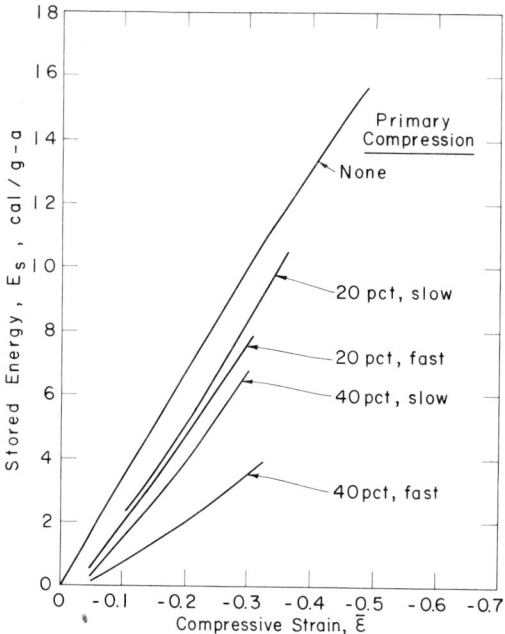

FIG. 19. The energy stored in copper in primary compression (upper curve) and in secondary compression after different amounts of slow or fast primary compression. Studenok.[122]

The effect of strain rate may be deduced from the results of Studenok.[122] He measured the energy stored in specimens of copper which had been subjected to primary compression of various amounts, namely 0, 10, 20, and 40% ($\bar{\varepsilon} = 0, -0.11, -0.23, -0.51$). The primary compression was performed at two different rates, reported as 0.05 and 2.5 mm/sec. The secondary deformation was by compression at 0.05 mm/sec. A selection of Studenok's results, re-plotted from the original graph with a change of units, is shown in Fig. 19. Although the absolute values may not be accurate, the difference between the effects of slow and fast primary deformation was established. A specimen deformed at a fast rate in the primary deformation stored less energy in the secondary deformation than did a specimen deformed at a slow rate in the primary deformation. This suggests that more energy was stored in fast than in slow primary deformation. According to an alternative interpretation, the observed effect was due to a change in the heat losses resulting from varying the time required for deformation, which may have left little or no true strain-rate effect.

3.1.4. *The Deformation Process*

The dependence of the stored energy on the extent and rate of deformation has been discussed in subsections 3.1.1 and 3.1.3. In many deformation processes both the extent and rate of deformation vary from volume element to volume element of the body being deformed. Such variations range from moderate, as in wire drawing, to severe, as in drilling. Also, as noted in subsection 3.1.3, the ratio of surface to volume of the deformed specimen may affect the amount of stored energy. For these reasons, when the stored energy is plotted as a function of a single parameter, such as the largest normal strain, the resulting curves may be expected to be different for different deformation processes.

The energy stored in a 75Au–25Ag alloy deformed by various processes is plotted in Fig. 7, which shows that the stored energy is not a unique function of the expended energy. In the deformation processes of tension, compression, and torsion, however, the stored energy plotted as a function of expended energy is a fairly close approximation to a single curve. This may be deduced from Fig. 10, which presents results obtained by Clarebrough et al.[46] with copper and nickel. In the processes of tension, compression, and torsion, no energy is lost in friction and the strain rates are usually low. The strain gradient in torsion does not have an appreciable effect presumably because of the predominance of the volume near the surface.

The data presented in Fig. 7 show that the largest values of the stored energy differ from one process to another. These values are for a 75Au–25Ag alloy. Corresponding differences between drilling[135] and wire drawing[128] have been observed with an 82.6Au–17.4Ag alloy. Fig. 17, which presents the strain-rate dependence of the stored energy in drawn wire of this alloy, shows that the highest level of the stored energy in drawn wires (43 cal/gram-atom) falls far short of the energy stored in drillings (89 cal/gram-atom). Whether this difference between the stored energies is accounted for solely by the difference in local temperatures during deformation, or whether other factors related to the nature and severity of the deformation processes play a significant role is still unknown.

Extremely severe rates of deformation are attained by shock loading. Appleton, Dieter and Bever[150] measured the energy stored in an explosively loaded 82.6Au–17.4Ag alloy. The energy increased with increasing shock pressure. The highest value was 31 cal/gram-atom for a pressure of 510 kbar. This value was slightly larger than the largest value of approximately 27 cal/gram-atom observed by Titchener and Bever[128] in wires of the same alloy drawn at moderate strain rates. At transient strains equal to the strains in wire drawing, or even at the same values of the hardness, shock loading caused more energy to be stored than wire drawing. The permanent strains, measured as reduction in thickness after shock-loading, did not exceed 5% and, as would be expected, did not correlate with the stored energies.

Iyer and Gordon[151] investigated the kinetics of the release of the stored energy from shock-loaded copper of two purities, but did not report absolute values of the energy. Scattergood, Beardmore and Bever[152] obtained values of the energy stored in an 82.6Au–17.4Ag alloy deformed by shock loading at two shock pressures; they were in good agreement with those of Appleton et al.[150] The kinetics of energy release in the investigations of Iyer and Gordon and Scattergood et al. will be considered in Part 4.

Beardmore, Holtzman and Bever[153] investigated the effects of shock loading on initially ordered and initially disordered specimens of the alloy Cu_3Au. The stored energy is shown as a function of shock pressure in Fig. 20. In agreement with the results obtained

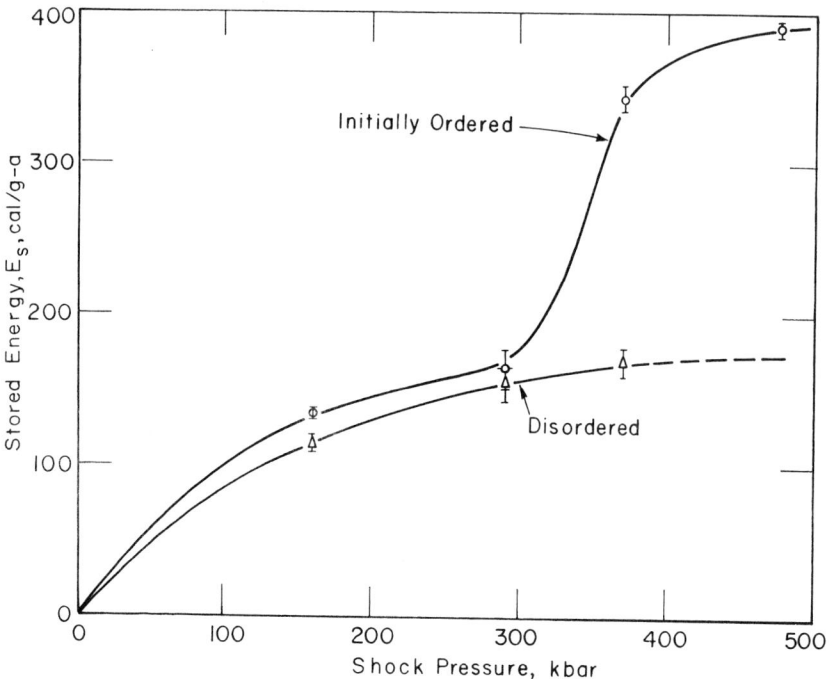

FIG. 20. The stored energy as a function of shock pressure in shock-loaded initially ordered and initially disordered Cu_3Au. Beardmore, Holtzman and Bever.[153]

with the 82.6Au–17.4Ag alloy, large amounts of energy were stored at small permanent strains. At a transient strain equal to the strain in cold rolling, shock-loaded Cu_3Au, ordered and disordered, stored more energy than was found by Cohen and Bever[130] in cold-rolled Cu_3Au.

Brillhart, De Angelis, Preban, Cohen and Gordon[154] measured the stored energy of oxygen-free high-conductivity copper shock loaded at pressures of 345 and 435 kbar. The stored energy was determined with an isothermal calorimeter described by Gordon.[104, 155] They found stored energies of 11.7 and 13.2 cal/gram-atom at 345 and 435 kbar, respectively; the respective transient strains were 0.22 and 0.26. The stored energies were higher than would be expected in copper strained slowly to these strains.

3.1.5. *Deformation History*

With a given deformation process, the effects of successive stages of deformation are generally assumed to be additive. Such an assumption is implied when the stored energy is investigated for processes such as wire drawing through more than one die. The same assumption is made in applications of single-step methods of measuring the stored energy in which deformation is carried out in successive stages. No investigator using a single-step method, however, appears to have investigated whether the energy stored after two stages of deformation is equal to that stored after deformation to the same total strain in a single stage.

The factors which could cause a difference between the energy stored in incremental and continuous deformation may involve the deformation mechanism, the transient energy release and recovery effects. Whereas in processes such as compression, tension, and torsion, the deformation mechanism should not be affected by intermittent operation, in processes such as wire drawing or rolling, the mechanism of deformation, in particular the redundant and frictional work, depend on the magnitude of each deformation step. In principle, these effects should alter both the expended and the stored energy, but apparently the effect on the latter is in general quite small. The effect of the transient energy release as observed by Williams[38, 43, 51–54] and Ham[39] may cause the stored energy to be smaller in a process carried out in several steps than in a process carried out in a single step. The most important cause of a difference between the energies stored in incremental and continuous deformation may be recovery occurring between deformation steps. The following analysis of the data presented by Studenok[122] will throw some light on this matter.

In Studenok's investigation, a specimen of copper was given a primary compression, presumably uninterrupted, to a strain $\bar{\varepsilon} = -0.225$ and after a secondary compression of $\bar{\varepsilon} = -0.150$ at the same rate stored 3.7 cal/gram-atom. Another specimen compressed to $\bar{\varepsilon} = -0.225$ in six stages stored 4.8 cal/gram-atom after an additional compression of $\bar{\varepsilon} = -0.150$. It may be concluded that the specimen given an uninterrupted primary deformation must have stored more energy in the primary deformation than the specimen deformed to the same primary strain with six interruptions. Thus, increasing the number of stages of compression appeared to reduce the amount of energy stored. The difference, amounting to about 9% of the stored energy in the example, was presumably lost by recovery between stages of deformation. An alternative interpretation of Studenok's results is that they are not sufficiently precise to support these conclusions.

Studenok's investigation[122] was primarily concerned with another aspect of strain history. As shown in Fig. 19, he observed that less energy was stored in a secondary compression when the primary compression was fast then when it was slow. As noted in subsection 3.1.3, this suggests that the energy stored in the fast primary compression was higher than that stored in the slow one. The effect of changing the rate of primary deformation is complex, as is shown by Studenok's observation that no energy was absorbed in the first 2% of the secondary compression of specimens which had undergone fast primary compression of 40% ($\bar{\varepsilon} = -0.51$).

3.1.5.1. *The Bauschinger Effect*

Degtiarev[123] subjected copper specimens to various amounts of primary tensile deformation and measured the energy stored in a secondary compressive deformation as a function of strain. Selected results are shown in Fig. 21. The amount of energy stored in

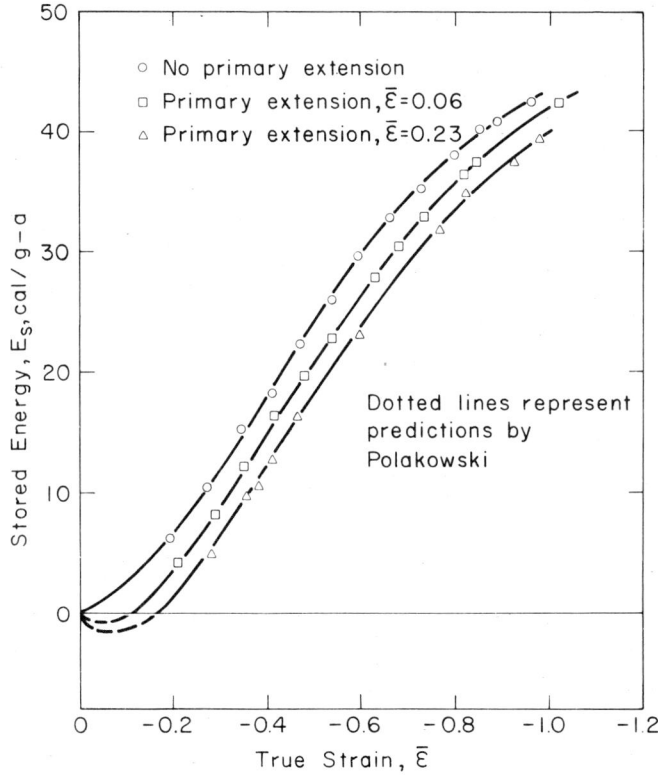

FIG. 21. Energy stored in secondary compression of copper after various amounts of primary extension. Degtiarev,[123] Polakowski.[156]

a secondary compressive deformation was reduced as the extent of the primary tensile deformation was increased. At low strains in the secondary compression, the amount of energy stored was reported to be too small to be measurable. Polakowski[156] concluded from Degtiarev's data that, at sufficiently small secondary strains, the stored energy decreased. Iyer and Gordon[157] confirmed this prediction with 99.999% copper as shown in Fig. 22. Their hardness measurements after reversed deformation also showed a Bauschinger effect.

Hargreaves, Loretto, Clarebrough and Segall[158] measured the energy stored in 99.99% copper subjected to 5 and 10% compression ($\bar{\varepsilon} = -0.05$ and -0.11) after 33% elongation ($\bar{\varepsilon} = 0.29$). They found that a 5% secondary compression reduced the stored energy from 9.0 to 5.8 cal/gram-atom; after 10% secondary compression the stored energy was 9.2 cal/gram-atom.

3.1.5.2. *Work Softening*

Titchener and Bever[159] investigated the effect of additional wire drawing at room

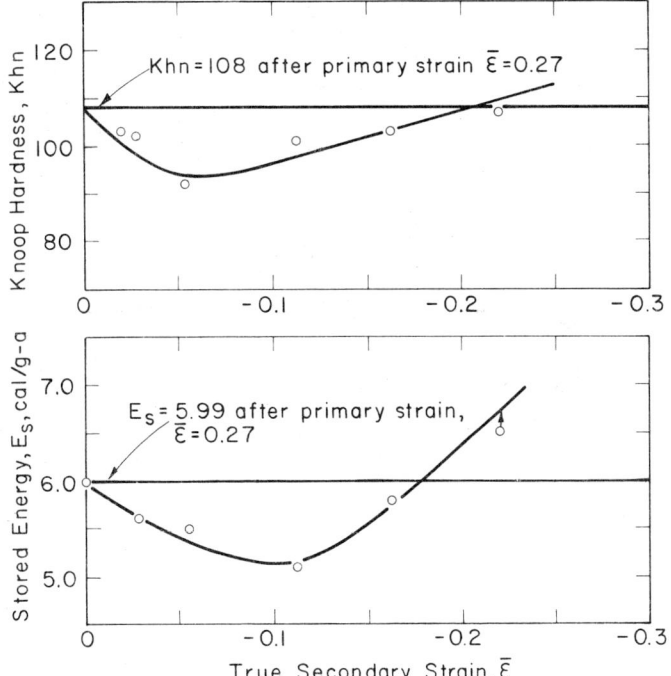

Fig. 22. Hardness and stored energy as functions of strain in secondary compression of copper after primary extension. Iyer and Gordon.[157]

temperature on the energy stored in an 82.6Au–17.4Ag alloy previously drawn at 78°K. The stored energy and the hardness decreased as shown in Fig. 23. In an investigation of the same alloy by Appleton and Bever[138] no work softening occurred during secondary torsion at 78°K of specimens subjected to primary torsion at 4.2°K. Instead, the stored energy increased with strain at a reduced rate until the values were equal to those obtained with specimens twisted solely at 78°K. In the same investigation, no effect on the stored energy measured at a strain of $nd/l = 0.6$ was observed when the strain rate was changed after straining to $nd/l = 0.3$ from $\dot{\gamma}_{max} = 0.3$ per min to $\dot{\gamma}_{max} = 0.2$ per min, although the stored energy at a strain of $nd/l = 0.6$ and a constant strain rate was a strong function of the strain rate over this range of strain rates (Fig. 18).

Erdmann and Jahoda[144] reported that the energy stored in a 92Pb–8In alloy, heavily cold worked at room temperature, decreased during the first 0.3% strain ($\bar{\varepsilon} = 0.003$) at 4.2°K.

Erdmann and Jahoda[145] also subjected a 70Cu–30Ni alloy to successive steps of tensile deformation, each step adding 0.5 to 2% strain, at increasing temperatures. The ratio of the stored energy to the energy expended in each step is plotted in Fig. 16. The minima in the curve are strong indications of an energy release by work softening. The general trend of the ratio falling with increasing temperature is, in part at least, associated with work softening. No hardness measurements were made.

3.1.6. *Cyclic Deformation (Fatigue)*

Welber and Webeler[80] made the first investigation of the stored energy after cyclic

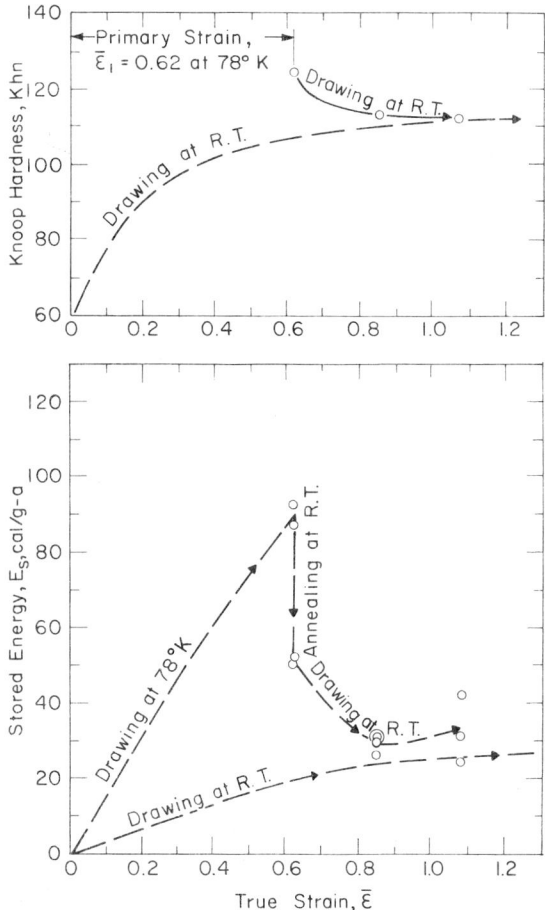

FIG. 23. Effect of strain and temperature history on the hardness and stored energy of an 82.6Au–17.4Ag alloy. Titchener and Bever.[159]

deformation. They reported values of the energy ranging from -23 to -32 cal/gram-atom for copper stressed in tension between 1000 and 20,000 psi for 4×10^6 cycles. Since negative values of the stored energy are impossible except in such special cases as after previous cold work, an explanation must be sought in experimental error. Welber and Webeler used an annealing method in which the stored energy was released in an initial run and a standard curve was established in a second run. Clarebrough, Hargreaves, Head and West[160] suggested that a change in the thermal resistance between the first and second heating may have taken place. This explanation requires that such a change occurred reproducibly in three experiments. Another explanation is that the specimens formed copper oxide at the surface during the cyclic straining and that during annealing endothermic dissociation of the oxide occurred with diffusion of some oxygen into the specimens.

Clarebrough et al.[160] found a small amount of energy stored in copper after cyclic straining in tension and compression. A more extensive investigation in the same laboratory obtained values of the energy stored in oxygen-free high-conductivity copper and 99.6%

nickel.[161] One series of copper specimens was strained cyclically in four-point reversed bending at a nominal stress amplitude of 25,000 psi until fracture, which occurred after approximately 2.8×10^4 cycles. A second series was strained cyclically in reversed tension at a stress amplitude of 10,000 psi until fracture after approximately 10^6 cycles. The values of the stored energy obtained in the two series of tests were 7.6 and 1.9 cal/gram-atom, respectively. The nickel specimens were cycled in reversed tension at a stress amplitude of 10,000 psi. The stored energy had the same value of 2.9 cal/gram-atom after 5×10^5 cycles and approximately 10^6 cycles. The kinetics of the release of the stored energy, which exhibited unusual features, will be discussed in Part 4.

Clarebrough, Hargreaves, West and Head[161] observed that cyclic deformation of oxygen-free high-conductivity copper, previously reduced 25% ($\bar{\varepsilon} = -0.29$) by rolling, caused work softening. Cycling was by four-point reversed bending at a stress amplitude of 21,000 psi until fracture after approximately 5×10^5 cycles. Although cycling produced an appreciable decrease in the hardness and changes in the kinetics of release of the stored energy, it did not appear to alter the amount of stored energy in the rolled specimen.

In several papers Dillon[61–63, 162] reported on the heat effect of cyclic torsion of aluminum tubes and rods. He also measured the work so that the stored energy could be calculated. There are differences between the stored energy values reported in the four papers. In the first paper describing the torsion of tubes;[61] a stored energy of 0.27 cal/gram-atom per cycle at a strain amplitude of 0.006 was reported. The ratio of the energy stored to the energy expended per cycle increased with strain amplitude from 6% at an amplitude of 0.001 to 28% at 0.006. A second paper on the torsion of tubes[63] reported that the energy stored was zero. The experimental apparatus had been refined for these later experiments, which suggests that the values reported in ref. 61 are incorrect. A stored energy of 0.07 cal/gram-atom per cycle at a strain amplitude of 0.003 was reported for aluminum rods cycled in torsion.[62] The ratio of the stored to the expended energy per cycle was 45%. In the experiments with rods the temperature rose to 260°C: as stated by the author, radial heat losses not accounted for would explain the apparent large values of the stored energy. In a later paper,[162] Dillon summed up his investigations by stating that except for the first two cycles starting with the fully annealed state, the fraction of work that appeared as heat was always between 95 and 100%. In Dillon's experiments the stress–strain hysteresis loops attained a steady state after two or three cycles. At that stage it was unlikely that much, if any, of the expended energy was stored. The stored energy values reported in refs. 61 and 62 should be regarded with reserve.

Dillon[64] also investigated cyclic straining in torsion of annealed copper tubes. From the measured heat and work, the stored energy could be calculated. Specimens from the same batch of copper tubes differed widely in the heat effect and the stored energy. The stored energy at a strain amplitude of 0.005 averaged 0.16 cal/gram-atom per cycle and the fraction of expended energy stored averaged 20%. The heat effect over one cycle was also examined. In some specimens over portions of a cycle, heat was observed to be generated at a faster rate than work was done, indicating a release of stored energy. Averaged over a cycle, however, the stored energy always increased.

Halford[69] subjected annealed and cold-worked oxygen-free high-conductivity copper tubes to cyclic deformation at constant strain ranges. He measured the stored energy by either a two-step or a one-step method. With the two-step method, he found that the annealed copper stored energy and the cold-worked copper released energy during cycling. The energy stored by the annealed copper specimens was 4.4 cal/gram-atom at 500 cycles and a strain range of 0.006, and 7.6 cal/gram-atom at 89 cycles and a strain range of 0.020. The stored energy of the cold-worked copper was reduced by 1.9 cal/gram-atom at 1000 cycles and a strain range of 0.008 and by 3.2 cal/gram-atom and 5.1 cal/gram-atom at 260 and 1000 cycles and strain ranges of 0.024 and 0.026, respectively. The kinetics of the release of the stored energy, which were similar to those observed by Clarebrough et al.,[161] will be discussed in Part 4.

Halford measured the change in stored energy over the first few cycles of deformation by a single-step method. The stored energy increased in annealed material and decreased in previously cold-worked material. The change was larger at larger strain amplitudes. The work of deformation was also measured. For annealed copper the ratio E_s/E_w decreased with increasing strain range and increasing number of cycles: at a strain range of 0.020, E_s/E_w was 15% after one half-cycle and 7.5% after 6 half-cycles. The slope of the curve of E_s/E_w plotted against number of cycles tended to zero as the number of cycles increased.

Halford also measured the change in stored energy during a half-cycle by a single-step method. Both annealed and cold-worked copper released stored energy immediately following each reversal of plastic strain. A curve showing the change in stored energy over three half-cycles of deformation at one strain range is presented in Fig. 24.

3.2. Variables Related to the Metal

3.2.1. *Grain Structure*

Clarebrough et al.[124] in the first investigation of the effect of the grain size on the stored energy deformed 99.98% copper by compression. At small strains, fine-grained specimens (average grain diameter ≈ 0.030 mm) stored more energy than coarse-grained specimens (average grain diameter ≈ 0.150 mm). At strains beyond $\bar{\varepsilon} \approx -0.4$, the difference decreased progressively. Above a strain of $\bar{\varepsilon} \approx -0.7$, at which the expended energy was approximately 380 cal/gram-atom, the energies stored in the fine- and coarse-grained specimens were the same within experimental precision ($\pm 5\%$). The results of Clarebrough et al. are shown in Fig. 25. Williams[163] criticized the curves of Clarebrough et al., but his criticism was refuted by the original authors[164] and by Titchener.[165]

Titchener and Bever[128] detected no effect of grain size in wires of 82.6Au–17.4Ag alloy drawn to a strain of $\bar{\varepsilon} = 0.58$. The grain sizes were 0.018 and 0.75 mm. This absence of an observable effect of grain size is not incompatible with the findings of Clarebrough et al.[124] on copper in which, at a strain of $\bar{\varepsilon} = -0.58$, the difference in stored energy was less than the precision of the measurements of Titchener and Bever.

Williams[51] reported values of the energy stored in 99.999% copper as a function of grain size. The deformation process consisted of compression by multiple impact to a maximum strain $\bar{\varepsilon} = -0.7$. The energy stored in specimens having a grain size of approximately 0.02 mm was about 15% larger than that stored in specimens having a grain size of approximately 0.6 mm. The difference in stored energy did not seem to disappear at large strains in contrast to the findings of Clarebrough et al.[124]

Loretto and A. J. White[166] observed an increase of over 30% in the energy stored in 99.999% copper deformed in compression to a strain $\bar{\varepsilon} = -0.36$ when the grain size was reduced from 0.700 mm to 0.150 mm. J. L. White[125] and J. L. White and Koyama[74] also found an increase in the stored energy with a decrease in grain size in oxygen-free high-

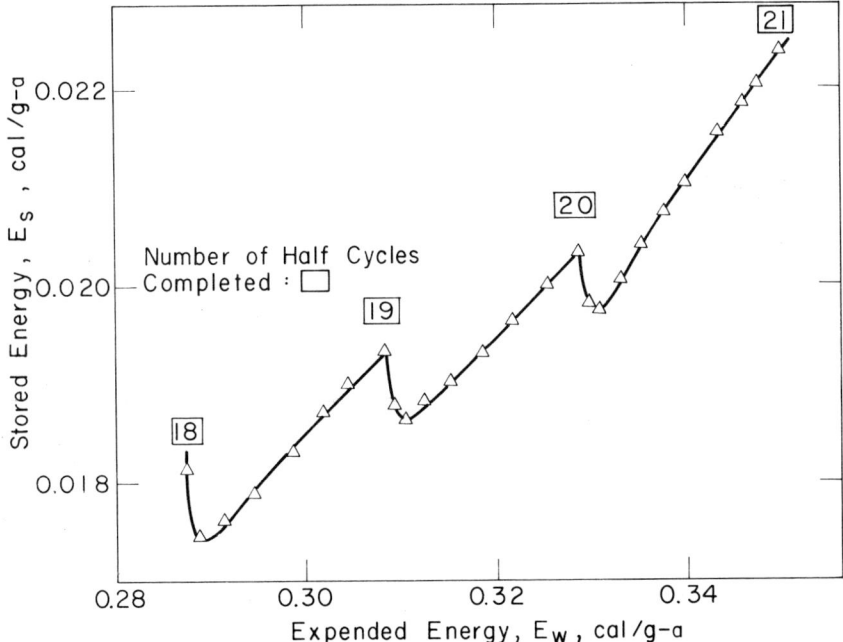

FIG. 24. Stored energy in copper versus expended energy over three half-cycles of reversed straining at a strain range of $\bar{\varepsilon} = 0.01$. Halford.[69]

conductivity copper and 99.999% copper both deformed by compression to strains of $\bar{\varepsilon} \approx -0.39$.

Williams[54] reported values of the energy stored in 99.999% copper of three different grain sizes deformed in tension to a maximum strain $\bar{\varepsilon} = 0.35$. The effect of grain size is shown in Fig. 26. At a strain $\bar{\varepsilon} = 0.3$ an increase in the grain size by a factor of approximately

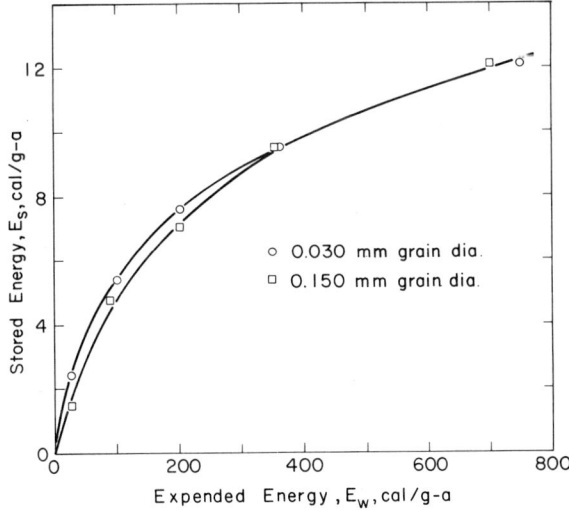

FIG. 25. The stored energy as a function of the expended energy in copper of two grain sizes. Clarebrough et al.[124]

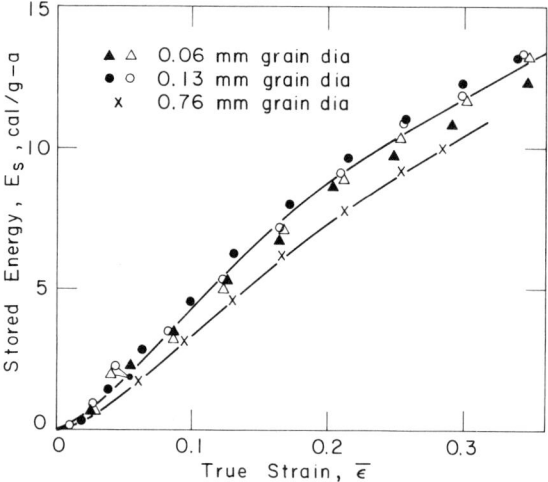

Fig 26. Stored energy as a function of strain in tension for 99.999% copper of various grains sizes. Williams.[54]

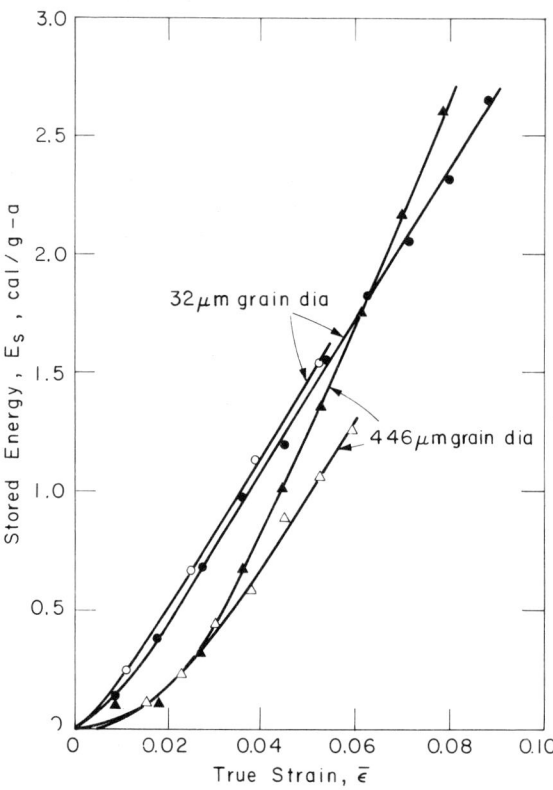

Fig. 27. Stored energy as a function of strain in tension for specimens of 82.6Au–17.4Ag of two grain sizes. Wolfenden.[59]

ten increased the stored energy by approximately 8%. There was no tendency for the curves to converge. The stress–strain curves also did not converge over the range of strains investigated.

Wolfenden[59] measured the stored energy of specimens of an 82.6Au–17.4Ag alloy deformed in tension to a maximum strain $\bar{\varepsilon} = 0.09$. The grain sizes were 32 μm and 446 μm. The results are given in Fig. 27. Up to a strain of $\bar{\varepsilon} = 0.06$, the fine-grained material stored more energy than the coarse-grained; above that strain the effect of grain size is not clear. The stress–strain curves converged at about a strain of 0.06. Wolfenden's values of the stored energy are compatible with those of Titchener and Bever.[128]

Preferred orientation may have an effect on the stored energy, as suggested by Williams.[51] Others have suggested this in discussing annealing textures[167] and recrystallization kinetics.[168] The stored energy is probably also affected by pre-existing substructure. The effects of neither preferred orientation nor substructure have yet been investigated, although results on single crystals can give some indication of the effect of preferred orientation.

3.2.1.1. Single Crystals

The first investigations of the energy stored by single crystals were those of Farren and Taylor[169] and Krivobok,[93] both published in 1925. Krivobok carried out thermal analyses of single crystals of an Fe–1.68Si alloy deformed by compression. He did not obtain values of the stored energy, but was able to show that it was a positive quantity. Farren and Taylor measured the energy in two single crystals of aluminum deformed by extension to strains ranging from $\bar{\varepsilon} = 0.14$ to 0.44. Neither the orientation nor the purity of the crystals was stated. At a strain of $\bar{\varepsilon} = 0.44$, the stored energy was 3.6 cal/gram-atom, which was less than the value these investigators obtained with polycrystalline aluminum extended to the same strain. Also, the ratio E_s/E_w of approximately 5% was smaller than that obtained with polycrystalline aluminum; the ratio was independent of strain.

Masima and Sachs[133] reported the ratio E_s/E_w for single crystals of brass of unstated composition deformed by extension, but gave no values of the stored energy. Kanzaki[79] with single crystals of copper deformed by compression, obtained values of the stored energy from 0.4 to 5.2 cal/gram-atom at shear strains ranging from 0.1 to 0.7. Although the strains cannot be directly compared, these values of the stored energy were less than one-half of the values reported by the same investigator using the same anisothermal annealing method on polycrystalline copper. As discussed in subsection 4.1.1, Kanzaki's values for copper should be treated with reserve.

Williams[50] measured the energy stored in two single crystals of 99.999% copper deformed by extension. The results are presented as a function of shear strain in Fig. 28. The stored energy plotted in this manner was independent of initial orientation. The curve is concave upwards, which differs strikingly from curves of stored energy versus strain for polycrystalline copper (Fig. 8). Williams reported that the incremental rate of energy storage $\delta E_s/\delta E_w$ in single crystals passed through a maximum of about 10 to 15% at a shear strain of approximately 0.1, which contrasts with the monotonic decrease of this rate in polycrystalline copper.

Taoka et al.[97] cold rolled two Fe–2.74Si single crystals to a reduction of 70% ($\bar{\varepsilon} = -1.2$). One crystal was oriented with a $\{110\}$ plane parallel to the rolling plane and an $\langle 001 \rangle$ direction in the rolling direction. The second crystal was oriented with a $\{100\}$ plane parallel to the rolling plane and a $\langle 001 \rangle$ direction in the rolling direction. The $\{110\}$

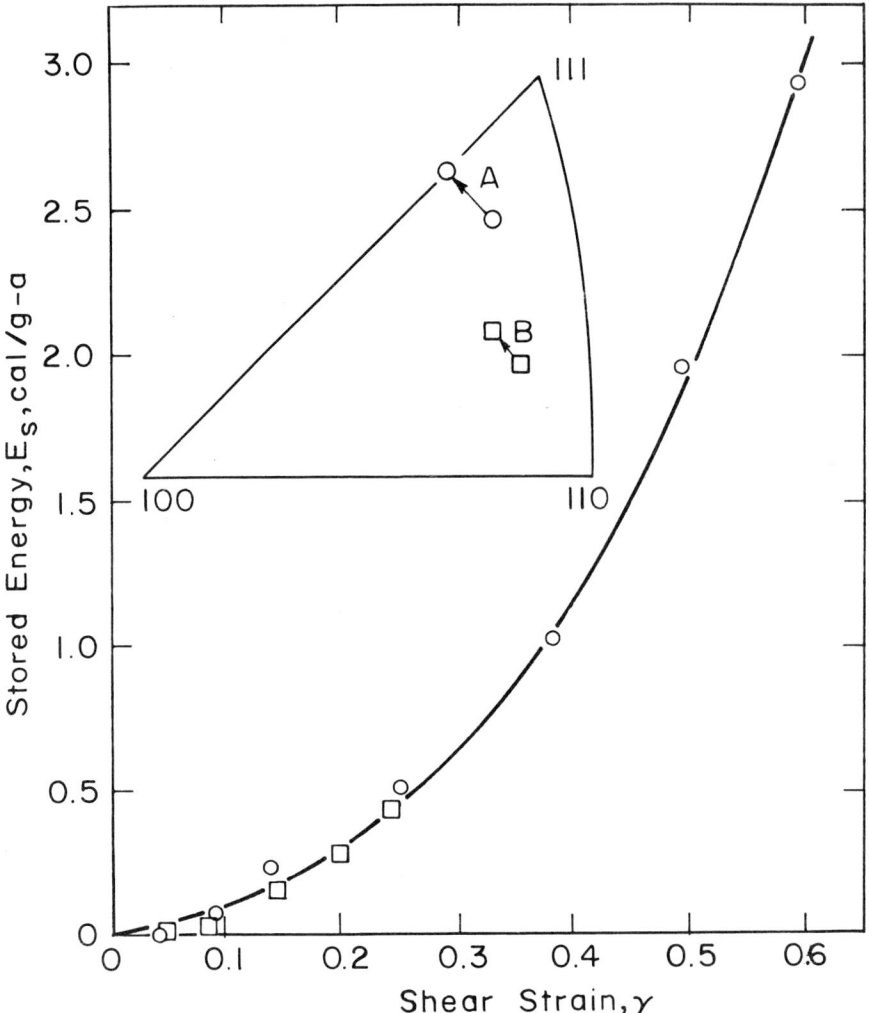

Fig. 28. Stored energy as a function of shear strain in two single crystals of copper. Orientations shown in stereographic triangle. Williams.[50]

⟨001⟩ crystal stored 394 cal/gram-atom and the {100} ⟨001⟩ crystal stored 210 cal/gram-atom. The kinetics of the release of the stored energy will be discussed in Part 4.

Nakada[68] measured the work and the heat effect of the compression of aluminum single crystals as a function of orientation and strain. Thus the stored energy can be computed from his reported data. Crystals oriented with the compression axis along [111] or [110] or along the direction through the center of the standard stereographic triangle (Schmid factor = 0.5) stored approximately the same fraction of the expended energy at the same shear strain. The ratio E_s/E_w decreased with strain, from 40% at $\gamma = 0.1$ to 15% at $\gamma = 0.75$. A crystal oriented with the compression axis along [100] anomalously stored much less energy. The stored energy as a function of shear strain is shown in Fig. 29. Crystals with [110] and [111] orientations have the same stored energy. The values are the

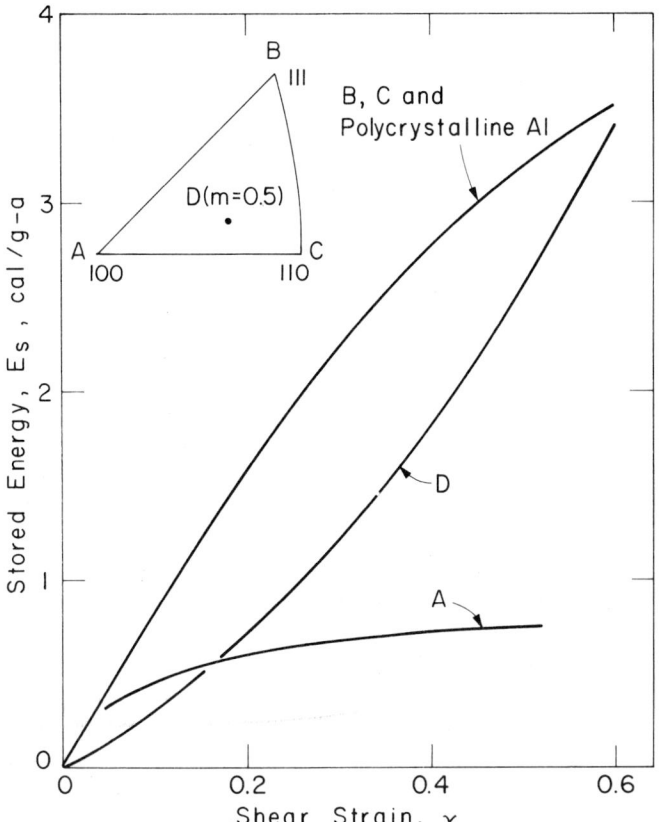

Fig. 29. Stored energy as a function of shear strain in single crystals of 99.99% aluminum compressed at various orientations. Data of Nakada.[68]

same as those of polycrystalline aluminum, which was also investigated. The stored energy increased from 0.15 cal/gram-atom at a shear strain of 0.05 to 3.6 cal/gram-atom at a strain of 0.6. The curve of the stored energy against shear strain for the crystal with a center-of-triangle orientation was concave upward as Williams found for copper single crystals.[50] Since E_s/E_w is not orientation dependent, the concave-upward shape of the curve must reflect the stress–strain curves of center-of-triangle crystals.

Nakada also investigated the heat effect of the compression of single crystals of silver as a function of orientation and strain. The ratio E_s/E_w was approximately the same for [111], [110] and center-of-triangle crystals at the same shear strain. The ratio E_s/E_w decreased with strain from 40% at $\gamma = 0.05$ to 20% at $\gamma = 0.35$. The stored energy as a function of strain is plotted in Fig. 30. The stored energy increased from 0.3 cal/gram-atom at a shear strain of 0.05 to 2.4 cal/gram-atom at a strain of 0.36. As for aluminum, the curve for the crystal with a center-of-triangle orientation was concave upward. At the same strain, the corner oriented crystals stored more energy than the center-of-triangle crystals. This reflects a difference in work-hardening rate.

Wolfenden and Appleton[146] and Wolfenden[147] used a single-step method to measure the energy stored in copper and aluminum single crystals strained in tension at 78°K. They

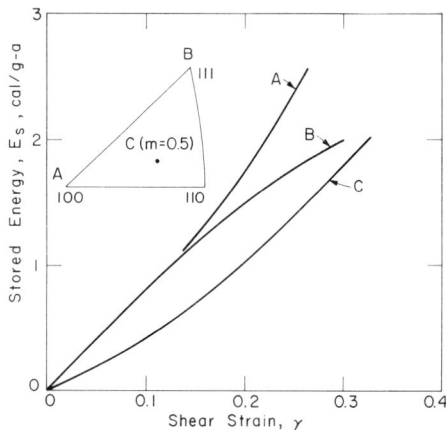

FIG. 30. Stored energy as a function of shear strain in single crystals of 99.99% silver compressed at various orientations. Data of Nakada.[68]

varied the strain and orientation and obtained curves of the stored energy as a function of tensile strain, which were generally concave upward over the range of strains investigated. In contrast the curves of stored energy against strain for polycrystalline copper and aluminum specimens, which were also reported, were concave downward as has generally been found with polycrystalline metals. A special feature of the results for the single crystals of copper and aluminum was that the energy stored by crystals of certain orientations was greater than the energy stored by polycrystalline specimens at the same strain, which is unlikely. Wolfenden and Appleton found no clear evidence for an orientation dependence of the stored energy. Nevertheless, they believed their results indicated that crystals with orientations close to [111] stored more energy than crystals oriented so that they had a distinct easy glide region.[146] On the other hand, Wolfenden[147] concluded from data in the form E_s/E_w vs. E_w that crystals oriented near [110] with a well defined easy glide region stored more energy than crystals of other orientations. These two conclusions are contradictory. The ratio E_s/E_w was reported to decrease with E_w from a maximum value of 100% at an expended energy E_w of 10 cal/gram-atom. The observations that the values of the stored energy in single crystals were greater than those in polycrystalline specimens, and that the ratio E_s/E_w was exceptionally high cast doubt on these results.

Wolfenden[55] measured the stored energy in copper single crystals strained in tension at room temperature. The curve of stored energy against shear strain was concave upward. Wolfenden concluded that his results indicated that crystals oriented with the tensile axis close to the [100] − [111] edge of the standard triangle stored more energy. This is not supported by the data. In a later paper Wolfenden[56] reported the ratio E_s/E_w and showed that there was an initial trend for E_s/E_w to increase with tensile strain which at larger strains tended to reverse. The maximum value of E_s/E_w reported was approximately 27%. A maximum in the curve of E_s/E_w vs. strain was most marked for crystals oriented for extensive easy glide. The trend of E_s/E_w vs. strain was similar to that found by Williams for $\delta E_s/\delta E_w$ vs. strain. The occurrence of a maximum was not reported by Wolfenden and Appleton[146] or Wolfenden.[147]

Wolfenden[57] measured the stored energy of silver single crystals strained in tension at room temperature. Over the range of strains investigated all but one of the curves of E_s

against shear strain were concave upward. From the findings of Williams,[50] Nakada,[68] Wolfenden and Appleton,[146] and Wolfenden[55, 57] it must be concluded that this is normal for single crystals. The stored energy values were smaller than those of Bailey and Hirsch[107] and Williams[52] for polycrystalline silver deformed to the same strain. Although Wolfenden suggested an orientation dependence of the stored energy, the data do not seem to support this suggestion. In a separate paper, Wolfenden[58] reported values of the ratio E_s/E_w for four of the crystals of the earlier investigation. In three crystals E_s/E_w increased with strain over the strain range $\gamma = 0.03$ to 0.29. In one crystal E_s/E_w increased at first and then decreased; the maximum of E_s/E_w as a function of strain was 13.5%.

Wolfenden[60] investigated the stored energy of aluminum crystals strained in tension. The values were small especially in relation to the experimental errors. This finding contrasts with that of Nakada[68] and Farren and Taylor[169] who observed measurable energy storage. Differences in purity and strain rate may account for some of the difference.

In his work on single crystals Wolfenden also obtained stress–strain curves and related the values of the stored energy to the stress. These results will be referred to in section 5.5. The relevance of the single crystal results to work-hardening theory will be discussed in 6.3.7.1.

From the results of all investigations on single crystals, we conclude that curves of stored energy versus shear strain of crystals with center-of-triangle orientations deformed in tension or compression are concave upward at least to the strain $\gamma = 0.5$. More experiments, however, are needed to establish an orientation dependence of the stored energy. It is known only that corner-oriented crystals store more energy at the same strain than center-of-triangle orientations. The behavior of the aluminum crystal with $\langle 100 \rangle$ orientation is an exception.

The ratio E_s/E_w has received considerable attention in interpretative papers. It appears to be independent of orientation when comparisons are made at equal strains and, in particular, is the same for corner orientations and center-of-triangle orientations. The ratio is generally higher for single crystals than for polycrystalline specimens of the same metal, but the strains in the tests of the single crystals were smaller than the strains in the tests of the polycrystalline specimens. This observation is consistent with the usual decrease of E_s/E_w with increasing strain in polycrystalline metals. In single crystals, however, there is an indication that at very small strains, the ratio E_s/E_w and the incremental rate of storage $\delta E_s/\delta E_w$ may increase with increasing strain so that the curves of E_s/E_w and $\delta E_s/\delta E_w$ versus strain show maxima.

3.2.2. Purity

The investigation of Clarebrough et al.[46] in 1955 on copper demonstrated that the purity of the metal played an important role in determining the amount of the stored energy. Evidence of the importance of impurities has since accumulated for nickel,[46, 52, 170, 171] aluminum[52, 172] and iron,[52, 97] in addition to further data for copper.[74, 124, 158, 173, 174]

As is discussed in subsections 3.1.1 and 3.2.3, the spread of measured values of the stored energy for a given metal such as copper is large. Since this spread is probably due in part to experimental errors and in part to the effects of impurities, a comparison of results obtained in different laboratories may lead to erroneous conclusions. The effect of impurities can be explored only by considering the results of single investigations or of several investigations from a single laboratory.

Clarebrough et al.[46] found that at a torsional strain of $nd/l = 1.87$ copper containing 0.35% arsenic stored more energy than oxygen-free high-conductivity copper nominally 99.98% pure. At the same expended energy, the ratio E_s/E_w was higher for the copper of lower purity (Fig. 10).

The energy stored in 99.98% copper deformed by compression has been the subject of three comparable investigations by Clarebrough et al.[46, 124, 174] The results of these investigations are included in Fig. 31, which indicates that the strain dependence of the

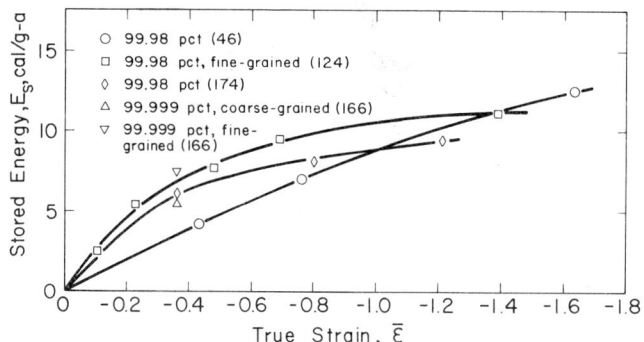

FIG. 31. Stored energy as a function of true strain in three batches of 99.98% copper and in 99.999% copper deformed by compression. Clarebrough et al.,[46, 124, 174] Loretto and White.[166]

stored energy may differ appreciably among specimens of copper of the same nominal purity. For example, the amounts of energy stored at 30% compression ($\bar{\varepsilon} = -0.36$) differ by as much as a factor of 2. Figure 31 includes points obtained in the same laboratory by Loretto and A. J. White[166] for 99.999% copper compressed 30% ($\bar{\varepsilon} = -0.36$). This material in an initially fine-grained condition stored more energy than any batch of 99.98% copper investigated in that laboratory. Further results from the same laboratory[46, 158] may be compared. As shown in Table 3.2, 99.99% copper extended 33% ($\bar{\varepsilon} = 0.29$) stored more energy than 99.98% copper and copper containing 0.35% arsenic extended to greater strains. Wenzl,[95] in an investigation of 99.999% and 99.98% copper and copper containing 0.3% arsenic, deformed by rolling, reported that the stored energy at a given strain increased slightly with increasing purity.

Table 3.2 includes results of J. L. White and Koyama[74] for copper deformed by compression. One lot ("Lot B") of oxygen-free high-conductivity copper stored 70% more

TABLE 3.2

The Effect of Purity on the Energy Stored in Copper

Investigators	Ref.	Grade of copper	Grain dia, mm	Deformation process	Strain $\bar{\varepsilon}$	Stored energy, E_s, cal/gram-atom	E_s/E_w × 100
Clarebrough et al.	46	99.55%	—	Tension	0.34	≈ 6.3	—
		99.98%	—	Tension	0.37	4.8	—
Hargreaves et al.	158	99.99%	—	Tension	0.29	9.0	—
White and Koyama	74	OFHC(A)	0.020	Compression	− 0.404	5.5	3.7
		OFHC(B)	0.020	Compression	− 0.392	9.3	5.7
		99.999%	0.020	Compression	− 0.352	6.7	4.7

energy than a second lot ("Lot A") of the same material having the same grain size and deformed to nearly the same strain. The ratio E_s/E_w was also higher. A specimen of 99.999% copper having the same initial grain size stored more energy than was stored by Lot A at a somewhat larger strain; also the ratio E_s/E_w was slightly higher.

We may conclude that the amount of energy stored by copper is not only a function of the total concentration of impurities, but also depends on the identity of the impurities and their distribution.

The effect of purity on the amount of stored energy in nickel can be inferred from investigations by Clarebrough et al.[46,170] and Williams.[52] As is evident from Table 3.3, the amounts of energy stored in two different batches of 99.6% nickel differed little. Appreciably more energy was stored by 99.6% nickel than by 99.85% nickel deformed to the same

TABLE 3.3

The Effect of Purity on the Energy Stored in Nickel

Investigators	Ref.	Composition, %	Deformation process	Strain, nd/l or $\bar{\varepsilon}$	Stored energy, E_s, cal/gram-atom
Clarebrough et al.	46	99.6 ("H")	Torsion	1.87	44.0
		99.6 ("K")	Torsion	1.87	35.2
Clarebrough et al.	170	99.85	Torsion	2.01	24.6
Clarebrough et al.	170	99.85	Compression	−1.21	17.6
		99.96	Compression	−1.21	18.5
Williams	52	99.38	Compression	−0.4	23
		99.9	Compression	−0.4	13

torsional strain. A further increase in purity from 99.85% to 99.96% appeared to increase the amount of stored energy slightly. Williams' results[52] for 99.38 and 99.9% nickel confirm the effect of impurities demonstrated by Clarebrough et al.[46,170] for 99.6 and 99.85% nickel.

Williams[52] measured the energy stored in 99.99% aluminum and in several batches of 99.9% aluminum deformed in compression by multiple impact. The values for 99.9% aluminum showed a large spread, for example, 6.3 to 9.9 cal/gram-atom at a strain $\bar{\varepsilon} = -0.4$. The values did not depend systematically on grain size. The energy stored in 99.99% aluminum fell within the range obtained with 99.9% aluminum.

Williams[52] also measured the energy stored in two different lots ("A" and "B") of ≈99.95% iron deformed in compression by multiple impact. From values of the energy at three different strains shown in Table 3.4 it may be inferred that the stored energy was affected by the normalizing treatment, and did not depend merely on grain size. Further, iron of Lot A stored about 50% more energy at a given strain than iron of Lot B given the same annealing treatment. This difference is likely to be mainly due to appreciable differences in carbon and oxygen contents. Lot A contained 0.03% carbon and 0.01% oxygen; Lot B 0.005% carbon and 0.04% oxygen. Table 3.4 also shows that the dependence of the stored energy on strain can differ appreciably. At low strains iron of Lot A air-cooled from 800°C stored less energy than iron of the same lot furnace-cooled from 1000°C; at high strains, the reverse held.

Taoka et al.[97] investigated the stored energy of two grades of iron, rolled to a reduction of 85% ($\bar{\varepsilon} = -1.9$). The carbon content of the first grade was 0.003% and that of the second

0.004%; the manganese content of the former was also lower. The first grade stored less energy than the second, namely 84 cal/gram-atom compared to 101 cal/gram-atom.

The results for nickel and copper already discussed indicate that a reduction in the overall purity below about 99.9% generally increases the stored energy. The results of Hertsriken et al.[96] for copper–zinc alloys containing 0.74 and 0.92% zinc and those of

TABLE 3.4

Energy Stored in $\approx 99.95\%$ Iron

Lot	Annealing temp, °C	Cooling	Grain dia, mm	Stored energy, E_s, cal/gram-atom at strain $\bar{\varepsilon}$		
				$\bar{\varepsilon} = -0.2$	$\bar{\varepsilon} = -0.3$	$\bar{\varepsilon} = -0.4$
A	800	Air	0.05	12	21	31
A	900	Air	0.05	10	17	—
A	1000	Furnace	0.02	15	21	27
B	1000	Furnace	0.10	9	13	—

Deformation: compression by multiple impact.
Ref.: Williams.[52]

Williams[51] for copper–silver alloys containing up to 1.0% silver support this generalization, but those of Wenzl[95] (p. 74) do not. When the concentration of impurities in copper falls below about 0.1%, the stored energy appears to become quite sensitive to variables other than the total concentration—presumably the identity and distribution of the impurity atoms. The complex relation between the stored energy and low levels of impurities make it unwise to base a generalization for nickel having a purity greater than 99.9% on a single investigation.[170]

The available evidence indicates that the identity and distribution of the impurity, suggested as the cause of some of the spread in the energy stored by copper purer than 99.9%, plays a similar role in aluminum purer than 99.9%. The same holds for 99.99% and purer iron and may apply to 99.9% and purer iron. In this respect, different face-centered and body-centered cubic metals appear to behave in the same manner and this may be true of all metals.

Impurities present in alloys in concentrations comparable to those in nearly pure metals may be expected not to have an appreciable effect on the stored energy. This is brought out by the results for gold–silver alloys obtained by Bever and co-workers in a series of investigations with different batches of material. Other solid–solution alloys, such as copper–zinc and copper–nickel alloys, may also be assumed to be insensitive to minor variations in the concentration of impurities. For this reason the values of the stored energy in alloys obtained in different investigations and by different investigators lend themselves to comparisons with fewer reservations than are indicated when values for nearly pure metals are compared.

3.2.3. *The Identity of the Metal*

The dependence of the stored energy on the extent of deformation in copper deformed at room temperature was shown in Fig. 2. The arguments developed in subsection 3.1.1 led to the conclusion that the high curves in Fig. 2 should be regarded with reserve. Even among the lower curves, however, an appreciable spread of values remains. The published values of the energy stored by other metals have similar spreads. This may be seen in Figs. 3 to 6, which show the stored energy after deformation at room temperature as a function of expended energy for nickel, iron, silver and aluminum.

For nickel (Fig. 3) it should be noted that the results of Bell and Krisement[109] do not agree with those of Clarebrough et al.[46] although Bell and Krisement drew a single curve through their values of the ratio E_s/E_w and those of Clarebrough et al. plotted against expended energy E_w. The two values of the ratio E_s/E_w reported by Bell and Krisement were obtained at values of E_w below the lowest used by Clarebrough et al.; the ratio changed rapidly at low strains.

The spread in the values of the stored energy reported for any given metal makes it difficult to compare the energies stored by different metals. Until the effects of all pertinent variables are known, it is advisable to restrict the comparison of the amount of energy stored by different metals to values obtained in the same laboratory and preferably by the same methods. With this restriction, few comparisons between the energies stored by different metals can be made.

Table 3.5 lists those investigations in which the effect of the metal as a variable can be seen. Investigations dealing with steel have been omitted from this table on the grounds

TABLE 3.5

Investigations in which the Metal is a Variable

Investigators	Ref.	Metals investigated†
Farren and Taylor	169	Cu, Al
Sato	73	Al, Ag, Cu, Fe
Rosenhain and Stott	41	Cu, Al
Maier and Anderson	15	Cu, Al
Fedorov	120	Cu, Al, Pb, Sn
Kunin	65	Cu, Ag, Cd
Hertsriken et al.	96	Fe, Cu
Clarebrough et al.	124, 170, 174, 175, 176	Ni, Ag, Cu, Au, Al
Williams	38, 51, 52, 54	Zr, Fe, Ag, Ni, Cu, Al, Pb
Schottky and Bever	177	Au, Ag
Smith and Bever	129	

† The metals listed for each group of investigators are in the order of decreasing stored energy at constant strain as determined by them.

that the effects of unknown differences in composition are likely to be too large to be ignored. The investigation of Quinney and Taylor[77] is not listed because the extent of deformation appears to have been different for each metal. The investigation of Panin and Milevskaya[70] has been omitted for the reason given in subsection 3.2.4 and that of Kanzaki[76] for the reason given in subsection 4.1.1.

Several investigations by Williams[38, 51, 52, 54] provide a set of comparable values of the stored energy for seven metals, which were deformed by multiple impact in the same calorimeter. The energy stored at a given strain decreased in the order zirconium, iron, silver, nickel, copper, aluminum, lead. With the exception of the position of silver, this is also the order of decreasing melting temperature, and Williams[52] presented the values of the stored energy at a strain $\bar{\varepsilon} = -0.30$ as a function of the ratio of the temperature of measurement to the melting temperature as shown in Fig. 32. Similar curves can be obtained from Williams' published values at other strains.

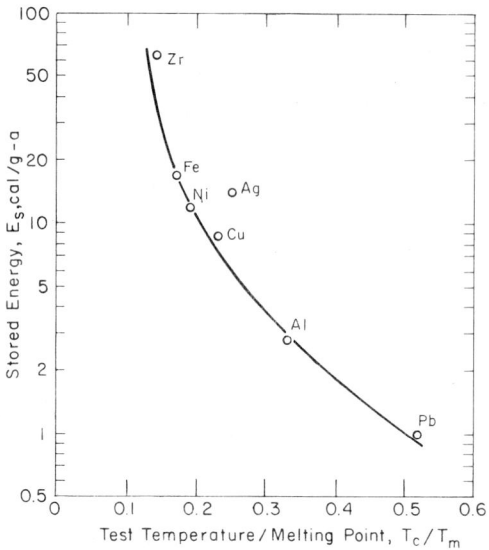

FIG. 32. Stored energy in various metals as a function of the ratio of test temperature to melting temperature T_c/T_m. Williams.[52]

The results of Clarebrough et al.[124, 170, 174–176] permit a comparison of the energy stored by five metals. The process of deformation and the method of measuring the stored energy were the same for all five, but the strain for copper and nickel was $\bar{\varepsilon} = -1.21$, compared with $\bar{\varepsilon} = -1.39$ for aluminum, silver and gold. The energy stored by nickel was approximately twice that stored by copper, as also observed by Williams.[38, 51, 52] The energies stored by silver, gold and aluminum decreased in that order. The energy stored by aluminum was about one-fifth that stored by silver, as was observed by Williams.[52] The stored energy may be estimated to increase by about 10% in copper and 15% in nickel when the strain is increased from $\bar{\varepsilon} = -1.21$ to $\bar{\varepsilon} = -1.39$. Increases of this magnitude do not change the order of the metals which, in the order of decreasing amounts of stored energy, is nickel, silver, copper, gold, aluminum. With the exception of the position of silver, this sequence agrees with that obtained by Williams (silver, nickel, copper, aluminum) and includes gold in the list. The generalization is confirmed that the order of decreasing stored energy is also the order of decreasing melting temperature, except for the position of silver. If the stored energy values of Clarebrough et al. are plotted against the ratio of temperature of deformation to melting temperature, as was done by Williams[52]

(Fig. 32), nickel, copper and aluminum fall on a straight line, but gold and silver do not lie on this line.

The order established from these two sets of investigations may be tested by means of other investigations listed in Table 3.5. The results of Farren and Taylor,[169] Rosenhain and Stott,[41] and Maier and Anderson[15] confirm the order copper, aluminum. Those of Fedorov[120] confirm the order copper, aluminum, lead, and indicate that tin stores about the same amount of energy as lead, as might be expected from their melting temperatures. Sato's results[73] confirm the order silver, copper, but do not confirm the position of iron or aluminum in the series. The results of Hertsriken et al.[96] confirm the order iron, copper. The order copper, silver obtained by Kunin[65] is the reverse of the order indicated by the work of Williams,[38,52] Clarebrough et al.,[174,176] and Sato,[73] which suggests that Kunin's results may be in error. The order gold, silver found by Bever and co-workers[177,129] is in the order of decreasing melting temperature.

The investigation of Khotkevich et al.[87] indicates that lead and cadmium store similar amounts of energy when deformed at 77°K. This suggests that cadmium should be placed with lead and tin, which would conform with the criterion of melting temperature. Also this grouping of cadmium is consistent with the observations of Kunin,[65] who obtained values for the energy stored in cadmium much lower than those in silver and copper. However, Khotkevich and Sirenko[178] in a later investigation found that cadmium compressed at 77°K stored over twice as much energy as lead deformed in the same manner.

In summary, the order established by the investigations of Williams, and Clarebrough and co-workers is generally confirmed, but there are minor exceptions. Since iron, zirconium, cadmium, and tin fall within a series that otherwise is face-centered cubic, it appears that the crystal structure of the metal is not important in determining the amount of the stored energy.

When the energy stored by different metals is compared at a given value of the expended energy instead of at a given strain, the differences among metals are greatly reduced. Table 3.6 lists values of the stored energy at two selected values of the expended energy for the

TABLE 3.6

Energy Stored by Different Metals at Constant Values of Expended Energy, E_w

Metal	Approximate stored energy, E_s, cal/gram-atom	
	at $E_w = 100$ cal/gram-atom	at $E_w = 400$ cal/gram-atom
Zr	19	59
Fe	6.5–8.5	18–25
Ag	11.5	24
Ni	8–11	15–24
Cu	7.5	14
Al	3–5	8–10
Pb	2	4

Ref.: Williams.[38,51,52,54]

The metals are listed in the order of decreasing stored energy at constant strain as determined in these investigations, all of which were conducted at room temperature.

seven metals investigated by Williams.[38, 51, 52, 54] At these two values of the expended energy, nickel, silver, copper and iron stored roughly equal amounts of energy. Aluminum stored less and lead much less than these four metals; zirconium stored much more, especially at larger values of the expended energy. Since the strength properties of different metals increase approximately with the melting temperature and, as has just been shown, the stored energy at a given strain also roughly parallels the melting temperature, a rough equality of the values of the energy stored in different metals at a given expended energy is not surprising.

3.2.4. *Alloys and Intermetallic Compounds*

Sato[73] in 1931 made the first measurements of the stored energy in alloys. He examined the dependence of the stored energy on composition in one alloy system. Since then, the energy stored in alloys has been investigated extensively. The alloys include terminal solid solutions, continuous solid solutions and intermetallic compounds; some of these had an order–disorder transition. Alloys with two or more phases have also received some attention.

Sato's data[73] show that, at the same expended energy, the energy stored in alpha brass deformed by torsion passed through a maximum at a composition of about 23% zinc, as may be seen in Fig. 33. Curves of similar shape are obtained if the stored energy at a given

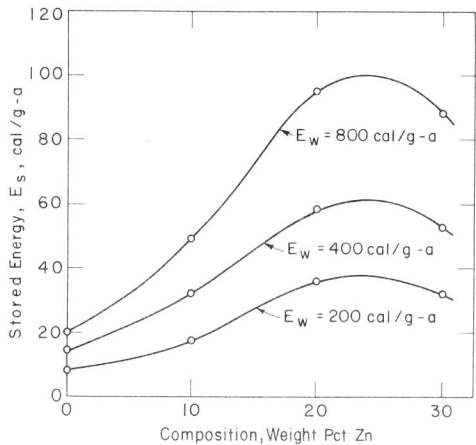

FIG. 33. Stored energy as a function of composition of α(Cu–Zn) alloys at various values of the energy expended in deformation by torsion. Data of Sato.[73]

torsional strain nd/l is plotted against composition. Sato found a value of 139 cal/gram-atom for the energy stored in a 70Cu–30Zn alloy twisted to a strain $nd/l = 1.83$. Clarebrough, Hargreaves and Loretto[179] measured a value of 153 cal/gram-atom at a strain $nd/l = 1.87$. Since also the shape of the curves of energy release presented by Sato for his annealing experiments closely resembles that of the curves presented by Clarebrough *et al.*[179] for a 70Cu–30Zn alloy, the dependence of the stored energy on the composition of alpha brass shown in Fig. 33 appears to be well established.

The energy stored in copper–zinc alloys of a single composition or of a narrow range of composition has been measured in several investigations.[53, 71, 77, 96, 179] These investigations contrast with that of Sato,[73] which covered an extensive range of composition. Quinney and Taylor[77] measured the stored energy in a 70Cu–30Zn alloy deformed in torsion to two strains. They found that the alloy stored more energy than copper. Williams[53] investigated the effect of annealing and quenching treatments on the energy stored in a 70Cu–30Zn alloy. He varied the short-range order of specimens by these treatments and measured their stored energy as a function of strain. In air-cooled specimens, the stored energy decreased and the incremental rate of energy storage $\delta E_s/\delta E_w$ increased with increasing annealing temperature above approximately 300°C. The curves for a strain $\bar{\varepsilon} = -0.30$, shown in Fig. 34, are typical.

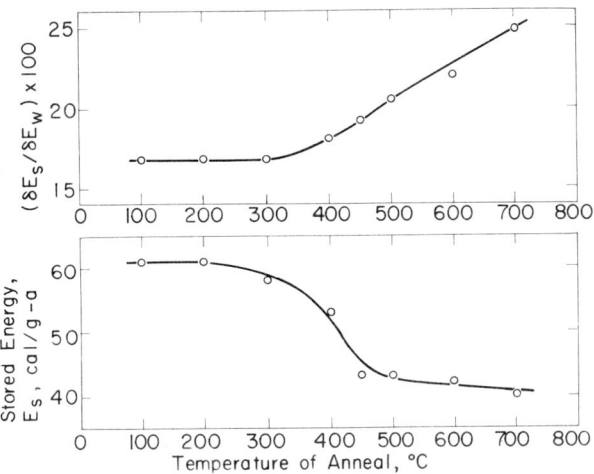

FIG. 34. Incremental rate of energy storage and stored energy as functions of temperature of prior annealing of a 70Cu–30Zn alloy subsequently deformed in compression to a strain $\bar{\varepsilon} = -0.30$. Williams.[53]

Hertsriken et al.[96] measured the energy stored in copper and copper alloys containing 0.74 and 0.92% zinc deformed in torsion. The energy stored at a given strain increased with increasing zinc content. The values of the stored energy appear to be in reasonable agreement with those of Sato,[73] but the kinetics of the release of the energy differ from those observed by Sato and Clarebrough et al.[179] (see Part 4). The precision of the measurements made by Hertsriken et al.[96] ranged from ±5 to ±25%; the ratio E_s/E_w for the alloy containing 0.74% zinc was unusually large.

Panin, Sukhovarov and Dudarev[71] measured the energy stored in copper–zinc alloys of two compositions deformed in tension. They observed that the energy stored by these alloys was greater than that stored by copper–nickel alloys at the same atomic concentrations deformed to the same strain.

Tizhnova[121] measured the energy stored by copper–nickel alloys over the composition range 0 to 70% nickel after deformation by compression. At a given strain, the stored energy increased sharply with increasing nickel content. Curves plotted from Tizhnova's data for specimens compressed 20 and 25% ($\bar{\varepsilon} = -0.23$ and -0.29) are shown in Fig. 35. According to Tizhnova, the ratio E_s/E_w also increased with increasing nickel content, but this increase was small in the range from 30 to 70% nickel.

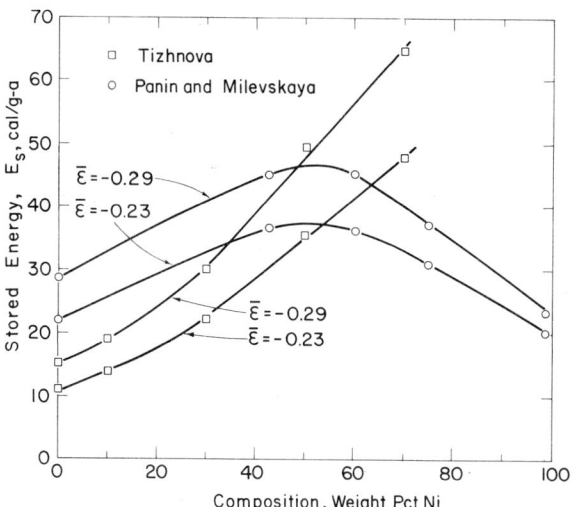

FIG. 35. Stored energy as a function of composition of copper–nickel alloys at two compressive strains. Panin and Milevskaya,[70] Tizhnova.[121]

Panin and Milevskaya[70] argued that the stored energy and the ratio E_s/E_w in copper–nickel alloys could not be monotonically increasing functions of the nickel content as reported by Tizhnova. They measured these quantities in 99.9% copper, in three alloys containing 54.2, 39.4, and 24.9% copper, and in 98.7% nickel deformed by compression. The curves obtained at 20 and 25% compression ($\bar{\varepsilon} = -0.23$ and -0.29) are included in Fig. 35. Curves of similar shape were obtained for 15 and 20% compression. Panin and Milevskaya also published relevant parts of the stress–strain curves for the alloys investigated and curves showing, as a function of composition, the ratio E_s/E_w at strains ranging from 15 to 30% compression. Their curves of E_s/E_w versus composition are inconsistent with the curves of the stored energy and the stress–strain curves.

The investigations of Tizhnova and of Panin and Milevskaya were made at the same laboratory and may, therefore, be expected to be comparable. The differences in their results revealed by Fig. 35, however, are striking. Tizhnova did not state the purity of the copper used in her investigation; Panin and Milevskaya used 99.9% copper. The values of the stored energy in copper obtained by the latter investigators are very high, even compared with those found in other investigations in the same laboratory.[71, 120–123] The values reported by Panin and Milevskaya for 98.7% nickel are lower than their values for 99.9% copper. Clarebrough et al.[46] found that 99.6% nickel stored about 50% more energy than 99.55% copper at the same torsional strain. The results of Sato[73] for copper–zinc alloys and Greenfield and Bever[136] for gold–silver alloys make it likely that the stored energy in the copper–nickel system also passes through a maximum, which probably lies on the nickel-rich side of the system. The existence of a maximum of the stored energy in the copper–nickel system still requires confirmation.

Erdmann and Jahoda[143a] also investigated the composition dependence of the energy stored in copper–nickel alloys. Since they deformed their specimens at 4.2°K, however, their results are not directly comparable with those of Tizhnova and of Panin and Milevskaya. The ratio E_s/E_w obtained by Erdmann and Jahoda is plotted against composition in Fig. 36. The ratio passes through a maximum at approximately 40% nickel. Erdmann and

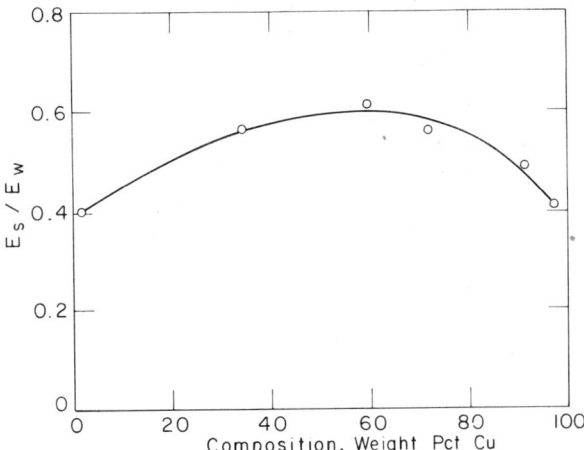

FIG. 36. The ratio E_s/E_w as a function of composition in copper–nickel alloys deformed in tension at 4.2°K. Erdmann and Jahoda.[143a]

Jahoda[143b] also presented stress–strain curves for 4.2°K and from these it is possible to estimate stored energy values. The curve of the stored energy at 10% strain against composition passed through a maximum at approximately 60–70% nickel and the extrapolated value for pure nickel was higher than that for pure copper.

Greenfield and Bever[136] investigated the composition dependence of the energy stored in chips prepared by drilling gold–silver alloys over the range 35 to 98 at. % gold. Drilling was carried out at room temperature and at 78°K. At both temperatures, the stored energy reached a maximum value at compositions near 50 at. % gold (Fig. 12). The plastic strains attained in drilling were unknown, but were certainly large. Since at large strains the stored energy is relatively insensitive to strain, the values of the stored energy for gold–silver alloys can be considered as values at constant strain and the trend of the stored energy against composition, therefore, can be compared with the trend observed by Sato with copper–zinc alloys (Fig. 33).

Panin et al.[71] found that the stored energy in terminal solid solutions of aluminum in copper increased rapidly with increasing aluminum concentration in dilute solutions. At higher concentrations, the increase was less rapid and at small strains there was a tendency for the stored energy to have a maximum at a composition lying between 4.5% and 7.5% aluminum. This tendency is similar to that found by Sato for the stored energy versus composition in alpha brass (Fig. 33).

Gangulee and Bever[180] measured the stored energy of silver-rich silver–magnesium solid solution alloys as a function of strain and magnesium concentration. The results for alloys with initial short-range order are shown in Fig. 37. The results for a 25.2 at. % magnesium alloy with initial long-range order are discussed in 3.2.4.1. Gangulee and Bever also measured the resistivity changes and correlated them with the stored energy values (see Part 5).

Waldman and Bever[181] measured the stored energy in a series of silver-rich silver–cadmium solid solution alloys at various strains and also measured resistivity changes (Part 5). They found that silver–cadmium alloys stored less energy than silver–magnesium alloys.[180] The resistivity changes, however, were larger.

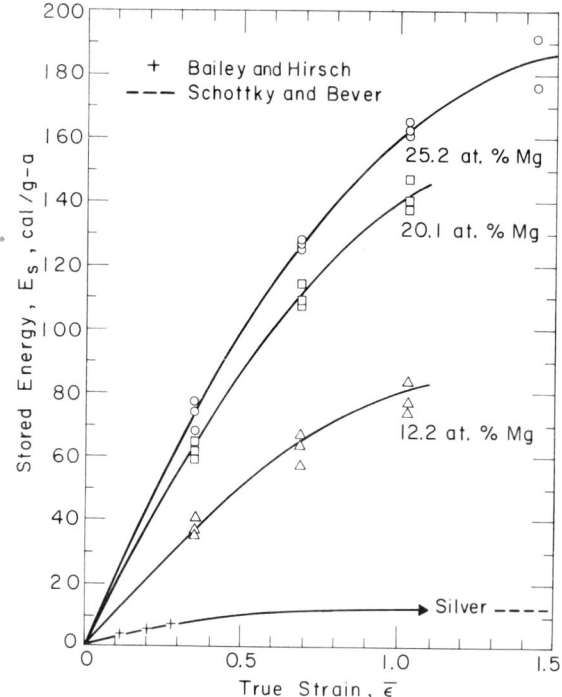

Fig. 37. The stored energy as a function of true strain in cold-worked silver–magnesium solid solution alloys with initial short-range order. Gangulee and Bever.[180] Data for silver presented for comparison. Bailey and Hirsch,[107] Schottky and Bever.[177]

The energy stored in dilute alloys of copper containing up to 1.0% silver was measured by Williams.[51] After compression by multiple impact to a strain $\bar{\varepsilon} = -0.3$, the energy stored in an alloy containing 1.0% silver was 11.5 cal/gram-atom compared with 9.5 cal/gram-atom stored in 99.999% copper. For silver concentrations up to 1.0% the stored energy appeared to be an approximately linear function of composition, but the data showed marked scatter. A small decrease in grain size accompanied the increase in silver content, but probably had only a minor effect on the stored energy.

Williams[53] measured the energy stored in a copper–aluminum alloy containing 6.7% (14.5 at.%) aluminum as a function of strain in compression by multiple impact after various heat treatments. Specimens slowly cooled from either 500° or 800°C stored more energy at a given strain than specimens quenched from these temperatures. When annealing treatments followed a quench, the amount of energy stored was intermediate between that stored in a quenched specimen and that stored in a slowly cooled specimen. Specimens quenched from 500°C stored more energy than specimens quenched from 800°C.

Van den Beukel[91] investigated the effects of 0.1 and 1.0% silver and gold in copper on the energy stored during compression at 77°K and released up to room temperature. Small concentrations of either element in copper did not appear to change the amount of energy released.

Taoka et al.[97] measured the stored energy of an Fe–2.98Si alloy reduced 85% by cold rolling ($\bar{\varepsilon} = -1.9$). The stored energy was 244 cal/gram-atom, which can be compared with

101 and 84 cal/gram-atom found for two grades of iron deformed to the same reduction in the same investigation.

Ham[39] investigated the energy stored in specimens of 99.99% copper and Cu–3.8Sn bent by a blow from a Charpy pendulum. While his main interest was in the transient energy release immediately after deformation, which will be considered in Part 4, he reported that the ratio E_s/E_w immediately after deformation was greater for the copper–tin alloy than the copper.

The curves in Figs. 12, 33 and 35–37 show that small changes in the composition of a binary solid–solution alloy do not greatly affect the stored energy. For this reason results obtained with different batches of an alloy with the same nominal composition are comparable even though there may be small accidental variations among them. This applies, for example, to the various investigations made with gold–silver alloys.[128, 135, 136, 138, 150, 159]

3.2.4.1. *Alloys with Long-range Order*

The energy stored by the deformation of the ordered alloy Cu_3Au has been the subject of several investigations.[130, 137, 153, 182] Cohen and Bever[130] measured the stored energy in initially long-range ordered and initially short-range ordered specimens as a function of strain by wire drawing and rolling. (Short-range ordered specimens have often been referred to as "disordered".) At any given strain, the energy stored by the specimens with initial long-range order was much greater than that stored by the specimens with initial short-range order, as shown in Fig. 38. Even in the short-range ordered alloy, however, the

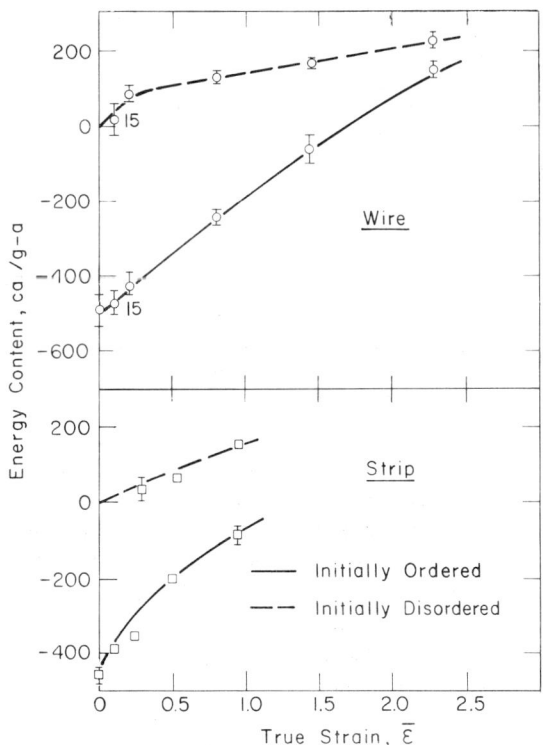

Fig. 38. Energy content as a function of true strain in initially ordered and initially disordered alloy Cu_3Au deformed by wire drawing and rolling at room temperature. Cohen and Bever.[130]

values of the stored energy were large and exceeded 200 cal/gram-atom at large strains. As shown by the figure, the energy contents of the materials in the two states tended to converge at large strains. It should be emphasized that the energy content of the initially long-range ordered alloy was lower than that of the initially short-range ordered alloy by an amount equal to the difference between the energy of long-range ordering and short-range ordering, which in turn is quite sensitive to the degree of short-range order present in the short-range ordered alloy. The ratio E_s/E_w was larger for initially short-range ordered than for initially long-range ordered drawn wires up to a strain $\bar{\varepsilon} = 0.75$. Beyond this strain, the reverse was true.

Roessler and Bever[137] measured the stored energy in wires of Cu_3Au drawn at 78°K. In both initially long-range ordered and initially short-range ordered specimens, this energy was much larger than that stored at room temperature, as may be seen in Fig. 39.

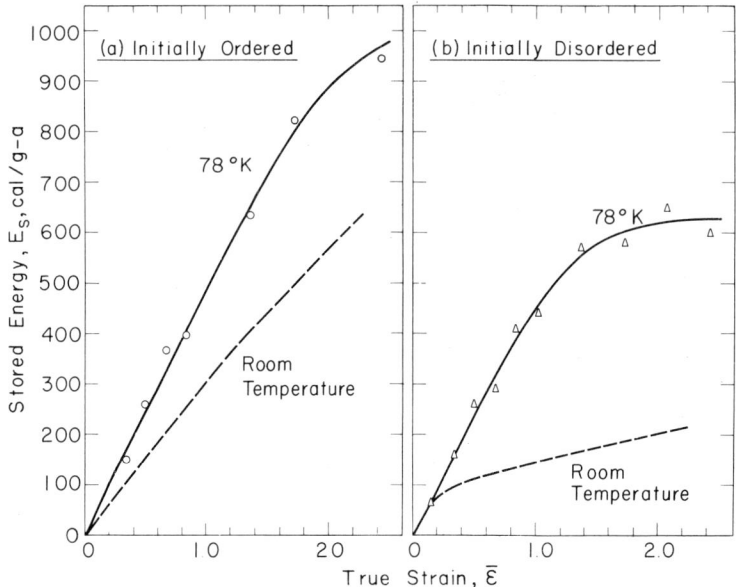

FIG. 39. Stored energy as a function of true strain in initially ordered and initially disordered alloy Cu_3Au deformed by wire drawing at 78°K. Roessler and Bever.[137] Values for wires drawn at room temperature also shown. Cohen and Bever.[130]

At small strains at 78°K, the rates of energy storage with respect to strain in specimens representing the two states of order were almost equal, which contrasted with the behavior at room temperature.[130] The difference in energy content between initially long-range ordered and initially short-range ordered specimens found by Roessler and Bever[137] was approximately 390 cal/gram-atom. At large strains at 78°K, as at room temperature,[130] therefore, the energy contents of the initially long-range ordered and initially short-range ordered alloys tended to converge. The ratio E_s/E_w was larger for initially short-range ordered than for initially long-range ordered wires up to $\bar{\varepsilon} \approx 1.0$. Beyond this strain, which corresponded approximately to the end of the linear dependence of the stored energy on strain in the initially short-range ordered wires (Fig. 39), the ratio was smaller for the short-range ordered than the long-range ordered wires. This is similar to the reversal of the relative magnitudes of E_s/E_w observed with initially short-range ordered and initially long-range ordered wires drawn at room temperature.[130]

Beardmore et al.[153] as mentioned in subsection 3.1.4 (Fig. 20) measured the energy stored in initially long-range ordered and initially short-range ordered specimens of the alloy Cu_3Au as a function of the shock pressure in shock loading. Up to a shock pressure of 290 kbar, the state of initial order had little effect on the amount of energy stored. Above 290 kbar, the energy stored by initially long-range ordered specimens increased sharply, whereas that stored by initially short-range ordered specimens increased only slowly.

Wolfenden[182] varied the degree of long-range order in Cu_3Au specimens by heat treatment. He did not measure the degree of order. The energy stored in tensile deformation was greater for the long-range ordered than the short-range ordered alloy. Alloys with intermediate degrees of order stored an intermediate amount of energy. Wolfenden also presented stress–strain curves. He suggested that the stored energy in Cu_3Au is proportional to the square of the flow stress, but the data do not support this suggestion and at present there is no theoretical justification for such a relationship in ordered alloys.

Robinson and Bever[140] measured the energy stored in silver–magnesium alloys containing 15.1, 18.1 and 18.5% Mg (44.2, 49.5 and 50.1 at.% Mg) as a function of torsional

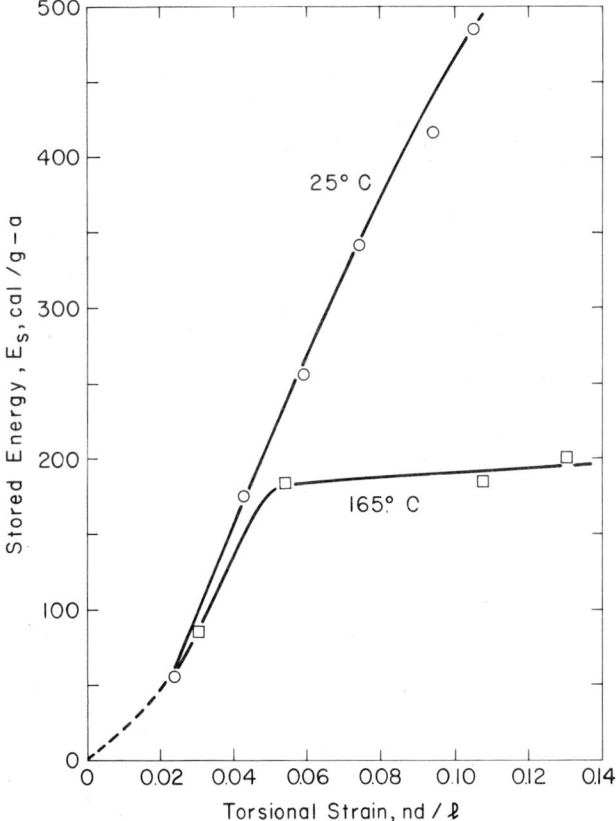

FIG. 40. Stored energy as a function of strain in a silver–magnesium alloy containing 44.2 at.% magnesium deformed by torsion at 25° and 165°C. Robinson and Bever.[140]

strain at room temperature and 165°C. The three alloys fall within the homogeneity range of the intermetallic compound AgMg, which has an ordered b.c.c. structure. The energy stored in each alloy increased linearly with strain at room temperature. At a strain of $nd/l = 0.052$, the alloys of the three compositions stored 226, 192 and 100 cal/gram-atom, respectively. The energy stored in the 44.2 and 50.1 at. % Mg alloys after deformation at 165°C reached plateaux of approximately 185 and 45 cal/gram-atom at torsional strains $nd/l \approx 0.050$ and 0.035, respectively. Results for the 44.2 at. % Mg alloy are shown in Fig. 40.

Robinson and Bever[183] measured the stored energy of the compound $TlBi_2$ deformed in torsion at room temperature, which is 0.6 of the absolute melting temperature. The stored energy was 5 cal/gram-atom at a strain of $nd/l = 0.06$, 20 cal/gram-atom at $nd/l = 0.31$ and 26 cal/gram-atom at $nd/l = 0.44$. These values are low in comparison with stored energies found in AgMg[140] and in Cu_3Au.[130]

The results of Gangulee and Bever[180] for the stored energy of an ordered silver-rich silver–magnesium alloy with composition near Ag_3Mg are presented in Fig. 41. The

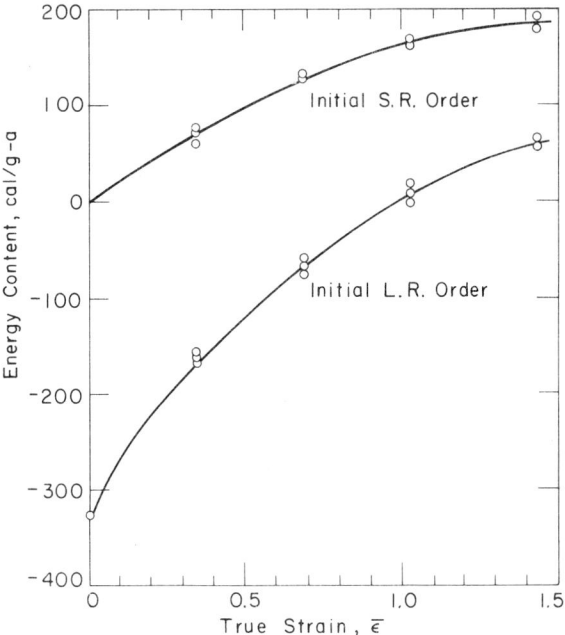

FIG. 41. The energy content as a function of strain in an alloy containing 25.2 at. % magnesium with initial short-range order and initial long-range order, relative to the unworked alloy with short-range order. Gangulee and Bever.[180]

energy stored by the alloy with short-range order is included. At the same strain the energy stored by the ordered alloy is much greater than that stored by the alloy with short-range order. The trends shown in Fig. 41 are similar to those in Fig. 38 for the alloy Cu_3Au.

3.2.4.2. *Two-phase Alloys*

Little work has been done on the stored energy of two-phase alloys. Khotkevich and Sirenko[178] measured the stored energy of lead–cadmium alloys deformed in compression at 77°K. The stored energy for strains of 20, 40 and 60% ($\bar{\varepsilon} = -0.22, -0.51$ and -0.92) as

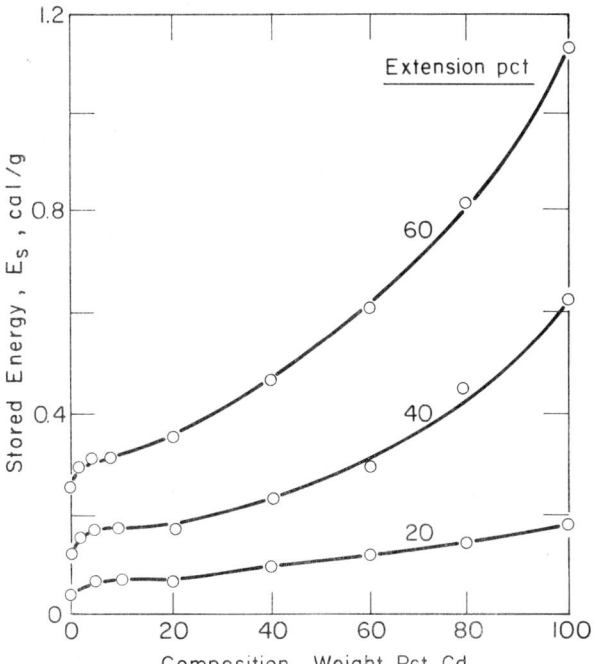

FIG. 42. Stored energy at three different strains as a function of composition of cadmium–lead alloys. Khotkevich and Sirenko.[178]

a function of composition is shown in Fig. 42. It can be seen that the dependence of the stored energy on composition does not obey a linear law of mixtures, which is difficult to explain. The ratio E_s/E_w was a function of composition as well as strain: it decreased in the order Cd, 60Cd–40Pb, Pb.

Chin and Grant[84,85] measured the stored energy in copper containing particles of Al_2O_3 in concentrations of 0.3 and 1.1% by volume after extrusion at 760°C. The stored energies ($\approx$ 26 cal/gram-atom) were larger than would have been expected for pure copper. This supported an interpretation of the good creep resistance of dispersion-hardened alloys involving high dislocation densities around the dispersed particles.

More work on multiphase alloys is desirable.

4. THE EFFECT OF VARIABLES ON THE RELEASE OF THE STORED ENERGY

When the temperature of a cold-worked metal is raised sufficiently, the stored energy is released. The calorimetric methods employing this kind of release to measure the total amount of stored energy also give information on the kinetics of the release. A knowledge of the kinetics of the energy release contributes to an understanding of the restoration processes of recovery and recrystallization, as well as to the interpretation of the mechanisms of energy storage.

The release of the stored energy has been investigated by anisothermal and isothermal annealing methods. The results obtained by anisothermal investigations are commonly presented in the form of a curve of the rate of release of energy ("power difference" between the cold-worked and the standard specimen) plotted against temperature. If the temperature is raised at a constant rate, the area under the curve is proportional to the amount of energy released. The curves of the power difference usually exhibit one or several peaks, each of which represents a surge in the release of stored energy. The peak occurring at the highest temperature corresponds to recrystallization. Peaks occurring at lower temperatures represent recovery processes, but recovery may also occur without a surge of energy release.

The results of isothermal annealing investigations are presented in the form of curves of power release versus time of annealing. As in anisothermal investigations, the area under the curve is proportional to the energy released. The shape of the curve depends on whether the temperature of annealing is high enough to cause recrystallization. In this case a characteristic peak is observed. If only recovery occurs, the rate of energy release falls continuously with time.

The release of the stored energy can also be investigated by annealing specimens for equal times at different temperatures, and measuring the amount of stored energy released during or retained after each annealing treatment. This method, which has been termed isochronal annealing, provides information on the amount of stored energy released as a function of annealing temperature, but not on the kinetics of the release during each annealing treatment. As a result the resolution obtainable is limited.

Anisothermal, isothermal, and a combination of isothermal and isochronal annealing data are amenable to analysis and may yield activation energies which characterize the mechanisms of recovery and recrystallization. The formal kinetics of the processes of recovery and recrystallization, and the methods of analyzing annealing data have been discussed in the literature.[184-186] The formal aspects of kinetics lie outside the scope of this review, but the interpretation of the results of analyses of the kinetics of the release of the stored energy will be discussed in section 6.4.

Those variables which determine the amount of energy stored in a cold-worked metal also affect the kinetics of its release. The discussion of the effects of these variables on the release of the stored energy is presented in sections 4.1 and 4.2 under the same headings as those of sections 3.1 and 3.2. Other variables relating to the annealing process are the rate of heating in anisothermal annealing, and the temperature of annealing in isothermal and isochronal annealing; they are discussed in section 4.3. The literature contains an extensive review by Clarebrough et al.[3] of the release of the stored energy.

The release of the stored energy has been investigated more often, and as a function of more variables, for copper than for any other metal. Nickel ranks next to copper in this regard. Section 4.1 and subsections 4.2.1 and 4.2.2 are mainly concerned with the results for these two metals. Other metals are given only limited consideration until subsection 4.2.3, where the identity of the metal is considered.

The amount of energy E_p released in prerecrystallization processes, that is, during recovery, and the amount E_r released during recrystallization, can be expressed as fractions of the total amount of energy stored E_s. The dependence of the ratios E_p/E_s and E_r/E_s on experimental variables will be dealt with in Part 4. In the literature the ratio E_p/E_r has also been used.

Some measurements of energy release immediately after deformation have been made in conjunction with the measurement of the stored energy by single-step methods. This energy release does not involve a separate annealing process. The measurements will be considered in section 4.4.

4.1. Variables Related to the Deformation Process

4.1.1. Extent of Deformation

Clarebrough et al.[46] observed by anisothermal annealing that the whole of the energy stored in copper of sufficiently high purity (99.98%), deformed at room temperature was released in a single peak, which corresponded to recrystallization. As the extent of deformation was increased, as shown in Fig. 43, the peak shifted to lower temperatures and the area under the peak increased, representing the increase in stored energy with increasing strain (subsection 3.1.1). Sato[73] reported similar observations.

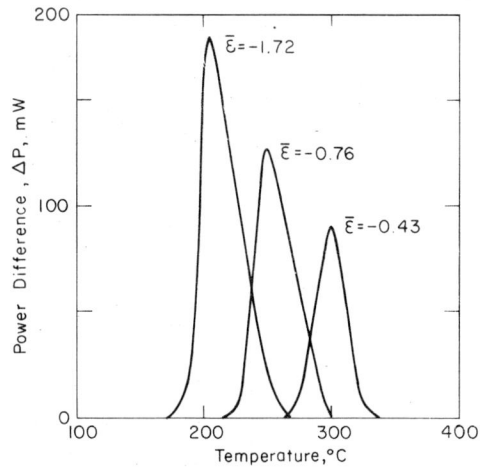

FIG. 43. The rate of energy release as a function of annealing temperature for 99.98% copper compressed to different strains. Clarebrough et al.[46]

Suzuki[14] and Kanzaki[76] in investigations of the kinetics of the release of energy by copper found two peaks. No other investigators have reported two peaks in annealing curves of copper deformed at room temperature. The amounts of stored energy reported by Suzuki and Kanzaki were exceptionally large; their observations of two peaks are judged to be of doubtful validity.

In an isothermal annealing investigation of 99.999% copper deformed in tension, Gordon[104] observed that, at a given annealing temperature, the recrystallization peak occurred at shorter times as the extent of deformation was increased (Fig. 44). He also observed that a small amount of energy E_p was released during recovery. The amount of this energy appeared to be independent of strain and the ratio E_p/E_s decreased from 10.8% at a strain $\bar{\varepsilon} = 0.10$ to 3.0% at a strain $\bar{\varepsilon} = 0.33$.

With 99.6% nickel deformed at room temperature Clarebrough et al.[46] observed by anisothermal annealing a recrystallization peak as in copper. This peak, designated C in Fig. 45, shifted to a lower temperature and enclosed a larger area as the extent of deformation was increased. Unlike 99.98% copper, 99.6% nickel, as shown in Fig. 45, released a large amount of energy during recovery. This energy release appears in the graph in part as an extended plateau and in part as a peak occurring at approximately 250°C. In a second batch of the same nominal composition, two peaks were observed during recovery, one at approximately 250°C, the other at approximately 530°C. The

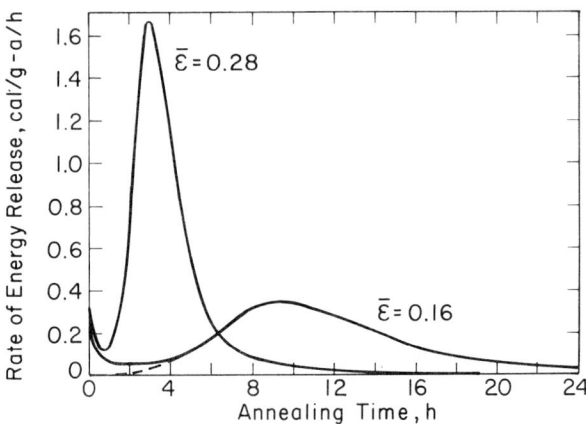

FIG. 44. The rate of energy release as a function of annealing time for 99.999% copper deformed by extension to two strains. In each case the isothermal annealing temperature was 189.7°C. Gordon.[104]

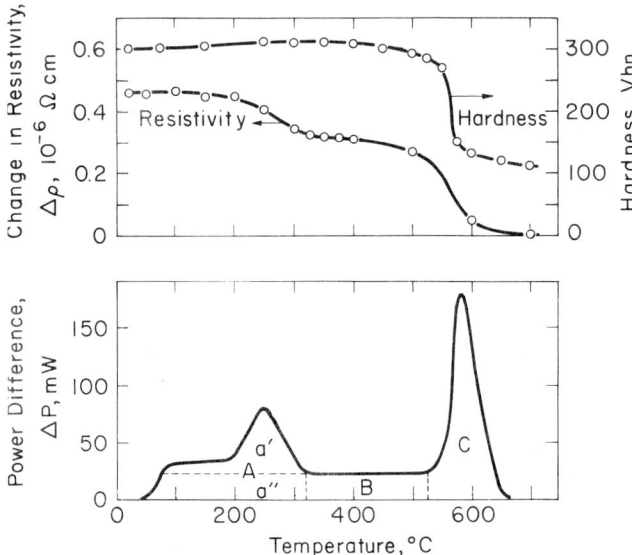

FIG. 45. The rate of energy release, hardness, and change in resistivity as functions of annealing temperature for 99.6% nickel deformed by torsion to $nd/l = 2.34$. Clarebrough et al.[46]

temperature range over which the plateau extended was not a function of the extent of deformation, nor did the recovery peaks shift with the extent of deformation. As shown in Fig. 46, the area under the plateau increased by approximately a third when the torsional strain was doubled from $nd/l = 1.0$ to 2.0. The area under the peak occurring at about 250°C increased only slightly with increasing strain. The ratio E_p/E_s decreased from 72 to 57% as the strain increased from $nd/l = 1.0$ to 2.0.

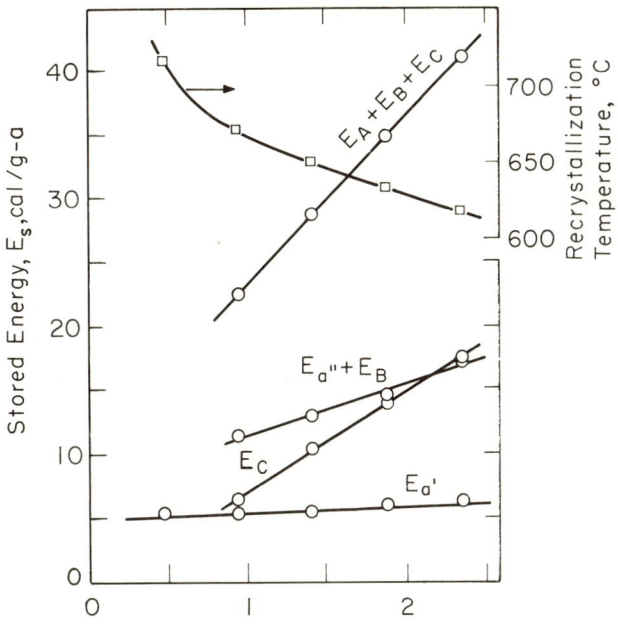

Fig. 46. Stored energy and recrystallization temperature as functions of strain in 99.6% nickel. Subscripts refer to areas indicated in Fig. 45. Clarebrough et al.[46]

Bell and Krisement,[109, 110] using isothermal calorimetry at several temperatures, measured the release of energy from 99.9% nickel deformed by compression at 213°K. They found two stages of energy release, one at 250°C and one at about 300°C; the latter was identified with recrystallization, while the former evidently corresponded to the first peak observed by Clarebrough et al.[46] in 99.6% nickel (Fig. 45). Bell[111] found stages of energy release at about 100°C and about 250°C in 99.9% nickel deformed at 213°K. He did not heat above 300 C.

The effect of the extent of deformation on the kinetics of the release of energy was also shown by results of Mima and Tokizawa[81] for aluminum and Kovacs[83] for copper. The results of Mima and Tokizawa are summarized in subsections 4.1.3 and 4.2.3.

4.1.2 Temperature of Deformation

When copper is annealed after deformation at 77°K, the spectrum of energy release includes a group of peaks below room temperature. Such peaks were first observed by Henderson and Koehler in 1956[88] for 99.999% copper. They found two main peaks in the vicinity of 180° and 250°K, and several others in the range 120° to 150°K. Loretto et al.[90] subsequently reported a complete spectrum including a recrystallization peak for 99.99% copper deformed by wire drawing at 77°K to a strain $\bar{\varepsilon} = 0.44$. Their results are shown in Fig. 47. The recrystallization temperature was approximately 30 C lower than that of a specimen of the same copper deformed to the same strain at room temperature. More energy was released during recrystallization, but recovery above room temperature appeared to release less energy from the specimens deformed at 77°K than at room temperature. The peaks below room temperature occurred in the vicinity of 140°, 180° and 250°K. Although the relative magnitudes of the various peaks shown by Loretto et al.

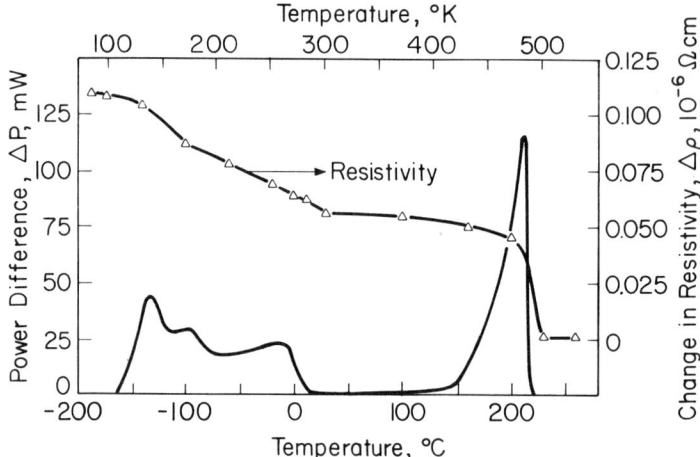

Fig. 47. The rate of energy release and change in resistivity as functions of annealing temperature in 99.99% copper deformed by wire drawing at 77°K. Loretto et al.[90]

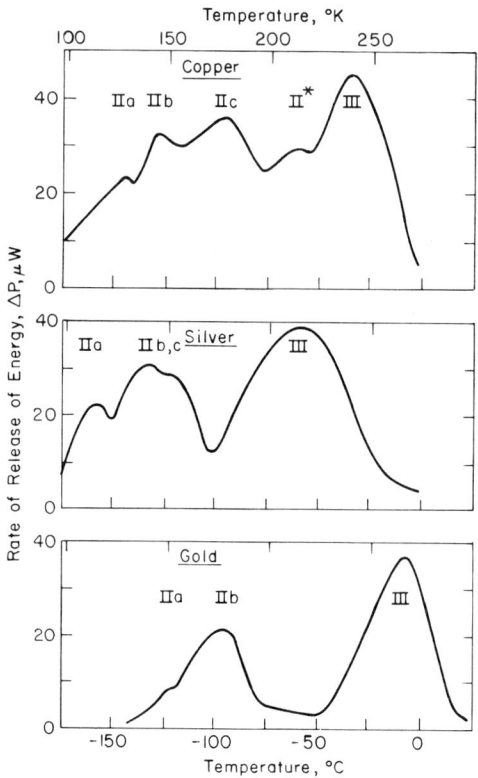

Fig. 48. The rate of energy release as a function of annealing temperature for copper, silver and gold deformed by compression at 77°K. Van den Beukel.[91]

differed from those reported by Henderson and Koehler, and although the observations of Loretto et al. did not resolve individual peaks at the lowest temperatures, the main features of the pattern of energy release found in the two investigations were the same.

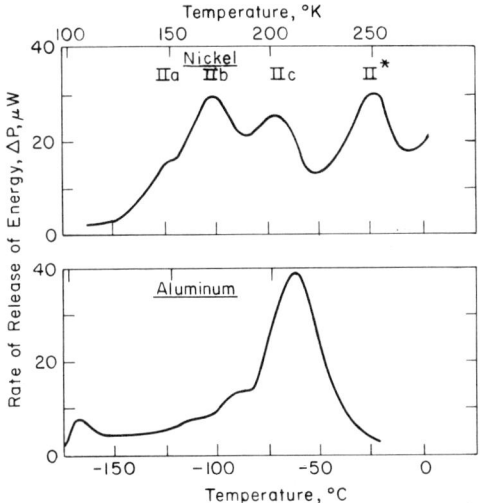

FIG. 49. The rate of energy release as a function of annealing temperature for nickel and aluminum deformed by compression at 77°K. Van den Beukel.[91]

The investigation of van den Beukel[89, 91] of the annealing of 99.999% copper from approximately 90°K to 30°C after compression at 77°K produced curves very similar to those of Henderson and Koehler.[88] Van den Beukel's results for copper are shown in Fig. 48.

The release of energy below room temperature in silver, gold, aluminum and nickel, deformed at 77°K, has also been investigated by van den Beukel.[89, 91] His results for silver and gold are included in Fig. 48; those for nickel and aluminum are shown in Fig. 49. The release spectrum for nickel was very similar to that for copper. As van den Beukel pointed out,[89] this similarity becomes more marked if the absolute temperature T_c of each peak or stage of energy release (identified by Roman numerals and letters in Fig. 48

TABLE 4.1
The Ratio of the Absolute Temperature of Annealing Peaks, T_c, to the Absolute Melting Temperature, T_m, of Copper and Nickel

Stage	T_c/T_m	
	Copper	Nickel
IIa	0.093	0.087
IIb	0.105	0.100
IIc	0.131	0.116
II*	0.155	0.145
III	0.178	—

Ref.: Van den Beukel.[89]

and Fig. 49), is expressed as a fraction of the absolute melting temperature (Table 4.1). Figure 49 shows that the rate of release of energy by nickel was appreciable at room temperature and rising; this suggests that the peak corresponding to Stage III occurs somewhat above room temperature. If the same value of the ratio T_c/T_m is assumed for nickel as for copper, Stage III in nickel may be predicted to occur at about 40°C. The peak observed by Clarebrough et al.[170] in 99.85% and 99.96% nickel at approximately 120°C after deformation at room temperature (see subsection 4.2.2) may, therefore, correspond to Stage III recovery.

Lugscheider and Wildhack[92] observed the release of energy below room temperature during anisothermal annealing of 99.99% aluminum deformed at 78°K. They found two stages of release, one in the temperature range 100° to 130°K and the other in the range 130° to 250°K.

Chuang and Bever[187] in an investigation of the annealing of 99.999% gold deformed at 78°K measured the energy retained after a 10 h anneal at 78°K, a 2 h anneal at 195°K, a 1 h anneal at 300°K and a 1 h anneal at 820°K. Major annealing stages occurred between these temperatures as was revealed by resistivity measurements. The relation between the stored energy changes and the resistivity changes occurring during annealing is discussed in Part 5.

The release of energy from alloys deformed at low temperatures is considered in subsection 4.2.4.

4.1.3. Rate of Deformation

An effect of strain rate on the kinetics of the release of energy was observed by Mima and Tokizawa.[81] They compressed aluminum of commercial purity slowly and by multiple impact. The temperature of the recrystallization peak was higher after slow compression than after compression by impact to the same strain. These results are considered in subsection 4.2.3.

The release of the energy by shock-loaded specimens will be a subject of the next subsection.

4.1.4. The Deformation Process

The dependence of the kinetics of energy release on the process of deformation has been observed in electrolytic copper (nominally 99.98%) by Clarebrough et al.[46] The recrystallization peak was broader when deformation was by torsion than when it was by tension or compression. This is not surprising since torsion produces a range of strains across the specimen. A similar broadening of the recrystallization peak of specimens deformed by torsion may be expected to occur in all metals.

The large amount of energy stored in nickel grindings has already been noted in subsection 3.1.3. The curves presented by Michell[188] and Michell and Haig[149] for the release of energy by grindings of 99.85% nickel indicate that most of this increase occurs in the energy released during recrystallization, E_r. This is consistent with the tendency of the ratio E_p/E_s to decrease with increasing strain as reported in subsection 4.1.1.

Iyer and Gordon[151] investigated the isothermal release of the stored energy after shock loading of 99.999% copper and tough-pitch copper. They showed that both materials released much of the stored energy before recrystallization. They compared the energy changes with changes in hardness. Results for 99.999% copper are presented in Fig. 50. In tough-pitch copper as much as 95% of the stored energy was released during recovery after shock loading, which should be compared with the smaller fractions after deformation by conventional processes.

Scattergood et al.[152] investigated the isothermal annealing kinetics of an 82.6Au–17.4 Ag alloy deformed by shock loading. They used hardness measurements to follow the course of the annealing and also made some measurements of the stored energy. The energy stored in a specimen loaded to 200 kbar decreased from 24 to 14 cal/gram-atom after annealing 5 min at 452°C. Recrystallization did not begin until after about 100 min. This

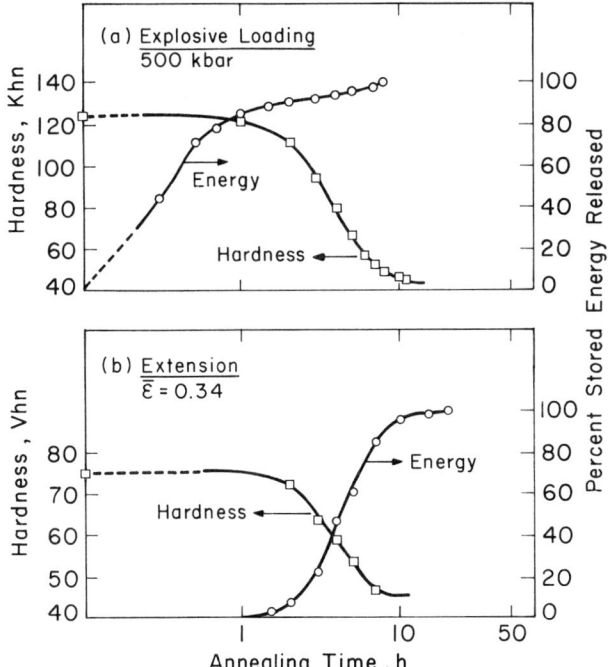

FIG. 50. Percent of energy released and hardness as functions of annealing time for copper deformed by (a) shock loading and (b) extension. Iyer and Gordon.[151]

release of a large amount of stored energy during recovery is in agreement with the findings of Iyer and Gordon for copper.[151] Caution should be exercised, however, in comparing the results for an alloy with those for a pure metal. As shown in subsection 4.2.4, the release of appreciable amounts of energy during recovery is generally observed in alloys deformed by conventional processes.

Brillhart et al.[154] found that approximately 90% of the energy stored in shock-loaded 99.9% copper was released in the last of three annealing stages. Major changes in resistivity and density also occurred in this stage. The release of the major part of the energy in a final annealing stage concomitantly with major changes in the resistivity and density is not consistent with the results of Iyer and Gordon. Brillhart et al.[154] reported, however, that the kinetics of the final annealing stage were typical of recovery rather than recrystallization.

The effect of the process of deformation on the release spectrum below room temperature has not been investigated. The unusual pattern of energy release after cyclic straining will be considered in subsection 4.1.6.

4.1.5. Deformation History

Hargreaves et al.[158] investigated the release of the stored energy after primary extension followed by secondary compression at room temperature. They used for this investigation a batch of 99.99% copper that was known to release appreciable amounts of energy during recovery after deformation. (The apparent contradiction of this statement with the first sentence in subsection 4.1.1 will be discussed in subsection 4.2.2.) Although the general form of the curve of power release versus temperature was only slightly altered as a result of the secondary compression, the amounts of energy released during recovery E_p and recrystallization E_r showed striking changes. The values of the ratio E_p/E_s fell from 57% after a primary extension of 33% ($\bar{\varepsilon} = 0.29$) to 45% after the same primary extension followed by a secondary compression of 5% ($\bar{\varepsilon} = -0.05$). Further secondary compression to 15% ($\bar{\varepsilon} = -0.17$) increased the ratio slightly to 48%. The values of the stored energy and the ratio E_p/E_s are shown in Table 4.2, which also lists the associated changes in density.

TABLE 4.2

Energy Released and Density Change during Recovery and Recrystallization of 99.99% Copper

Strain, $\bar{\varepsilon}$		Stored energy, cal/gram-atom			Ratio	Fractional change in density $\times 10^5$		
Primary extension	Secondary compression	E_p	E_r	E_s	$(E_p/E_s) \times 100$	$(\Delta d/d)_p$	$(\Delta d/d)_r$	$(\Delta d/d)$
0.29	0.0	5.1	3.9	9.0	57	2.1	5.6	7.7
0.29	−0.05	2.6	3.2	5.8	45	3.7	3.2	6.9
0.29	−0.17	4.4	4.8	9.2	48	2.9	6.3	9.2

Ref.: Hargreaves et al.[158]

4.1.6. *Cyclic Deformation (Fatigue)*

The results for cyclic straining of 99.98% copper obtained by Clarebrough et al.[161] are shown in Fig. 51. After cyclic reversed bending at a stress amplitude of 25,000 psi the energy was released in two broad peaks. The peak occurring at the higher temperature corresponded to localized recrystallization around fatigue cracks and at the surface of the specimen. After cyclic straining in reversed tension at a stress amplitude of 10,000 psi the whole of the energy was released in a single broad peak by a recovery process. This was in marked contrast to the usual behavior of the 99.98% copper used by these investigators, which

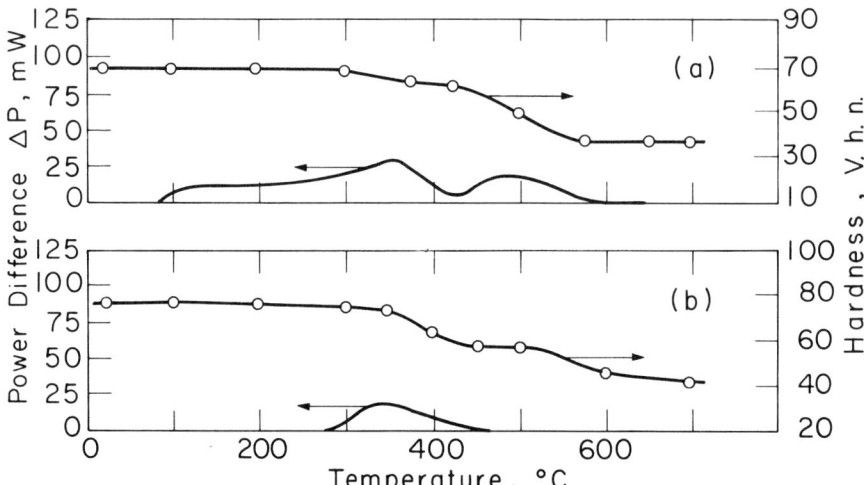

FIG. 51. Power difference and hardness for specimens of copper deformed by fatigue at a high stress (a), and fatigue at a low stress (b). The heating rate was 6°C/min. Clarebrough et al.[161]

when deformed unidirectionally released almost all of the stored energy in a single peak (Fig. 43) corresponding to recrystallization.[46]

Clarebrough et al.[161] also investigated the release of energy after cyclic loading of 99.6% nickel. They observed two broad peaks centered at approximately 300° and 530°C, but found no release of energy above 600°C. They considered the release to be incomplete, however, since the apparatus was not capable of detecting small changes in energy above 600°C. The power–time curves were independent of the number of cycles of stress and did not show the peak at 250°C which was found when 99.6% nickel was annealed after torsional deformation[46] (Fig. 45).

Halford[69] reported that the energy released by oxygen-free high-conductivity copper cycled in torsion over a strain range of $\gamma_{max} = 0.006$ occurred as a small peak whereas the energy stored in a specimen cycled over a range of $\gamma_{max} = 0.02$ was released over a temperature range of approximately 100°C. No trends in the kinetics of energy release from cyclically strained specimens after previous cold work were reported.

4.2. Variables Related to the Metal

4.2.1. Grain Structure

Clarebrough et al.[124] observed that an increase in grain size in 99.98% copper deformed by compression broadened the peak occurring during anisothermal annealing and shifted it to a higher temperature. An increase in the grain size of 99.999% copper caused a similar broadening and shift of the annealing peak.[166] In the earlier investigation,[124] the peaks obtained from the coarse-grained material remained broader and occurred at higher temperatures than those obtained from the fine-grained material, even at large strains at which the energies stored by initially fine-grained and initially coarse-grained specimens were the same (subsection 3.2.1).

J. L. White and Koyama[74] also observed that the recrystallization peak of energy release in oxygen-free high-conductivity copper and in 99.999% copper broadened and shifted to higher temperatures with increasing grain size.

4.2.1.1. Single Crystals

Kanzaki[79] compressed single crystals of copper of unstated purity at 93°K, 193°K and room temperature. His curves of energy release versus temperature for single crystals, obtained by anisothermal annealing, resembled his curves for polycrystalline copper[76] in having two peaks. They should be treated with reserve as discussed in subsection 4.1.1.

Taoka et al.[97] found that single crystals of Fe–2.74Si cold rolled to a reduction of 70% ($\bar{\varepsilon} = -1.2$) released energy in two stages during anisothermal annealing. The first stage occurred in the temperature range of 100° to 300°C, the second in the range 300° to 750°C. Polycrystalline Fe–2.98Si also released energy in two stages. Orientation affected the amount of energy released in the second stage more than the amount released in the first stage: a crystal oriented with an $\langle 001 \rangle$ direction in the rolling direction and a $\{100\}$ plane parallel to the rolling plane released 2.3 cal/g (125 cal/gram-atom) during the second stage and 1.56 cal/g (85 cal/gram-atom) during the first stage, whereas a crystal oriented with an $\langle 001 \rangle$ direction in the rolling direction, but with a $\{110\}$ plane parallel to the rolling plane released 6.3 cal/g (341 cal/gram-atom) in the second and 0.97 cal/g (53 cal/gram-atom) in the first stage.

4.2.2. Purity

The curves obtained by Clarebrough et al.[46] with a specimen of 99.98% electrolytic copper and shown in Fig. 43 were presented in subsection 4.1.1 as typical of the release of energy from high-purity copper annealed anisothermally. Electrolytic copper of slightly lower purity (99.96%) released a small but appreciable amount of energy during recovery. The results for specimens from the two batches of copper deformed to fracture in torsion are shown in Fig. 52. The release of energy by the 99.96% copper during recovery is

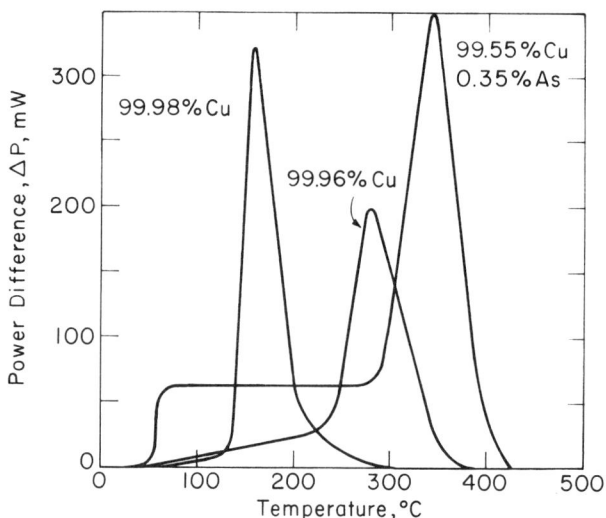

FIG. 52. The rate of energy release as a function of annealing temperature for copper of three different purities. Clarebrough et al.[302]

represented by the shoulder preceding the main peak. Results obtained from 99.55% copper containing 0.35% arsenic and deformed to fracture in torsion are included in Fig. 52. Copper of this composition released approximately half of the stored energy at a uniform rate from about 70°C until the beginning of recrystallization. The amounts of energy E_p and E_r associated with recovery and recrystallization increased with strain, as shown in Fig. 53. Wenzl,[95] in an investigation of copper of three different purities, observed no release of energy before the beginning of recrystallization.

The complex dependence of the stored energy on impurities, discussed in subsection 3.2.2, is reflected in the kinetics of the release of the energy. Hargreaves et al.[158] presented curves for the release of energy from 99.99% copper. These curves have an extended plateau, representing a uniform rate of release during recovery, followed by a recrystallization peak. They closely resemble the curves obtained by Clarebrough et al.[46] for arsenical copper containing about 99.55% copper, but are unlike curves for 99.98% copper. The curves of Loretto and A. J. White[166] for 99.999% copper deformed 30% in compression ($\bar{\varepsilon} = -0.36$) differed from those for 99.98% copper deformed in compression to various strains in showing small amounts of energy released by recovery. On the other hand, the curves presented by J. L. White and Koyama[74] for 99.999% copper did not indicate a recovery stage of energy release. These results show that the total concentration of impurities does not of itself determine the pattern of energy release.

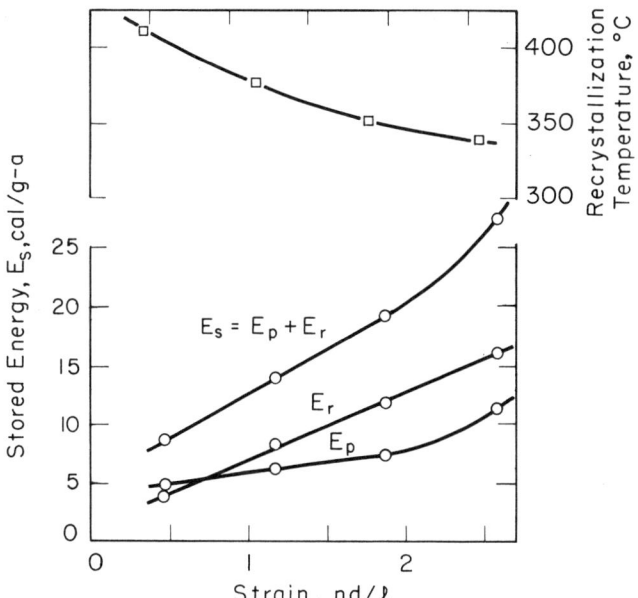

Fig. 53. Stored energy and recrystallization temperature as functions of strain in 99.55% copper containing arsenic. E_p and E_r refer to recovery and recrystallization. Clarebrough et al.[46]

The effect of purity on the pattern of energy release in nickel has been investigated by Clarebrough et al.[46, 170, 189] Results for one batch of 99.6% nickel are presented in Fig. 45. The curve of energy release includes a small recovery peak centered at about 250°C. A second batch of nickel of the same nominal purity showed two recovery peaks,

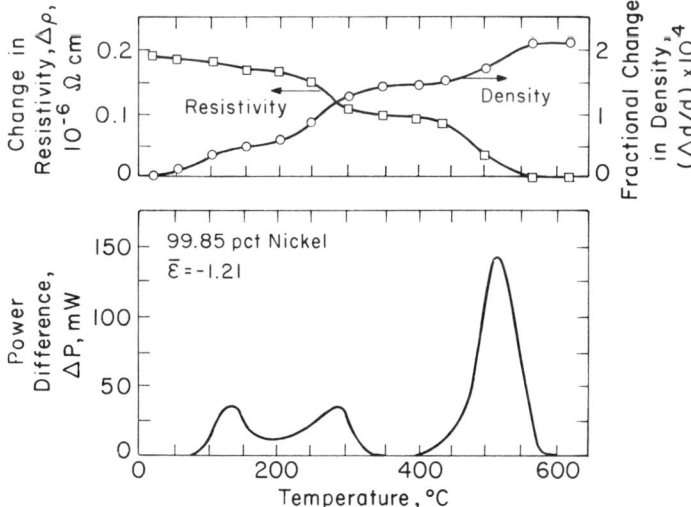

Fig. 54. Rate of energy release, change in resistivity and change in density as functions of annealing temperature for 99.85% nickel deformed by compression. Clarebrough et al.[170]

one at about 250°C, the other at about 530°C.[46] This striking difference indicates that in nickel, as in copper, the pattern of the release of the stored energy is not uniquely determined by the total concentration of impurities.

The release of energy by 99.6% nickel may be compared with the release by 99.85 and 99.96% nickel (Fig. 54 and Fig. 55). In nickel of the two latter compositions the extended

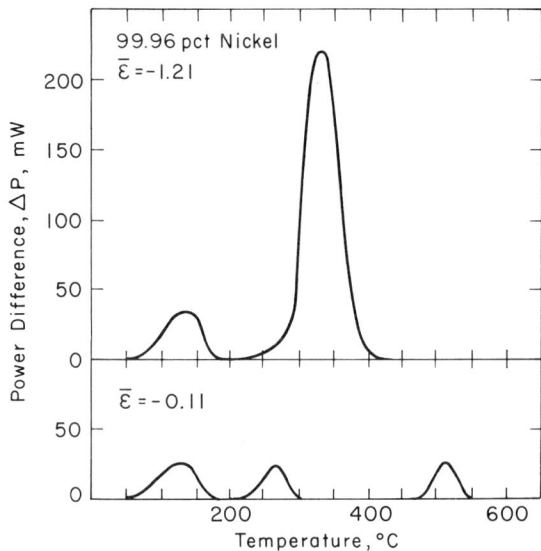

FIG. 55. Rate of energy release as a function of annealing temperature for 99.96% nickel deformed by compression to two strains. Clarebrough et al.[170]

plateau of energy release was absent. Instead, two peaks occurred at approximately 120° and 270°C. In the 99.96% nickel, the second peak was obscured by the recrystallization peak when deformation was severe.

Bell[112] found single peaks of energy release in recovery at 80°C in 99.999% nickel and at 250°C in 99.6% nickel. A specimen of intermediate purity (99.9% nickel) developed peaks both at 80° and 250°C.

The curves of energy release by two grades of iron cold rolled to 85% reduction ($\bar{\varepsilon} = -1.9$) obtained by Taoka et al.[97] are shown in Fig. 56. The iron of lower carbon content released energy over a lower temperature range even though it stored less energy than iron of higher carbon content.

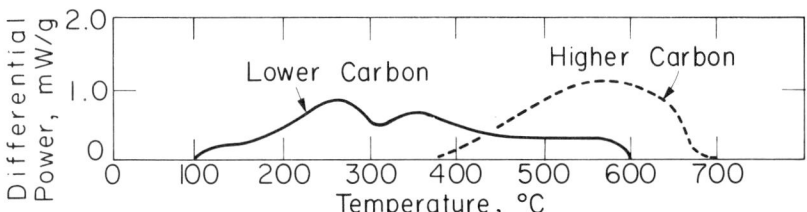

FIG. 56. Rate of energy release as a function of temperature in iron of two carbon contents cold rolled to an 85% reduction in thickness. Taoka et al.[97]

4.2.3. *The Identity of the Metal*

4.2.3.1. *Copper and Nickel*

In section 4.1 and subsections 4.2.1 and 4.2.2 the effects of variables on stored energy have been discussed primarily as observed with copper and nickel. The release of the stored energy by other metals will be discussed in this subsection.

4.2.3.2. *Silver and Gold*

Clarebrough et al.[176] observed that the energy stored in 99.98% silver and 99.9% gold deformed 75% in compression ($\bar{\varepsilon} = -1.39$) at room temperature was released in a single peak corresponding to recrystallization. In the much earlier investigation by Sato,[73] the energy stored in silver of undeclared composition deformed by torsion was released in a single peak, preceded by a small shoulder that may have represented a slight amount of recovery. The area under the peak increased and the temperature of the peak decreased with increasing extent of deformation. In agreement with the results of Clarebrough et al.,[46] the area under the shoulder of the curve increased with increasing deformation.

The curves of energy release obtained by Kanzaki[76] for 99.9% silver deformed by compression differed from those of Clarebrough et al.[176] and Sato[73] in having two peaks. Kanzaki's results for silver, like his results for copper, should therefore be treated with reserve.

Bailey and Hirsch[107] investigated the release of energy during isothermal annealing of 99.99% silver deformed by extension. In agreement with the results of Gordon[104] for 99.999% copper, the energy was released in two overlapping stages. In the first stage, representing recovery, the rate of release fell continuously with time. The second stage, representing recrystallization, produced the peak characteristic of this process. As was also observed by Gordon with copper, the ratio E_p/E_s decreased as the strain was increased. Kovacs[83] found that the energy stored by silver of unreported purity was released in a single peak at about 210°C.

Van den Beukel[91] investigated the release of energy below room temperature from 99.999% silver and 99.999% gold deformed by compression at 77°K (Fig. 48). He found complex annealing spectra. As with copper, the amounts of energy released in Stages II and III were roughly linear functions of the true strain. The annealing curves of the three metals had broad similarities but the spectrum increased in complexity in the order gold, silver, copper.

4.2.3.3. *Aluminum*

Quinney and Taylor,[77] in their account of an investigation of aluminum by anisothermal annealing, gave no information on the kinetics of the energy release. The general patterns of release of energy from aluminum observed by Kanzaki[76] and Clarebrough et al.[175] are in fair agreement, but the amounts of energy reported by Kanzaki were much larger than those reported by Clarebrough et al. The latter investigators used 99.991% aluminum compressed 75% ($\bar{\varepsilon} = -1.39$). Their results are shown in Fig. 57. Approximately 35% of the total stored energy was released in a long, slowly rising plateau extending from 120°C to the beginning of recrystallization at approximately 270°C.

Mima and Tokizawa[81] measured the energy released during the anisothermal annealing of commercially pure aluminum after slow compression and compression by repeated impact to various strains. Specimens compressed to large strains released energy in three overlapping peaks. This finding is compatible with those obtained from the isothermal

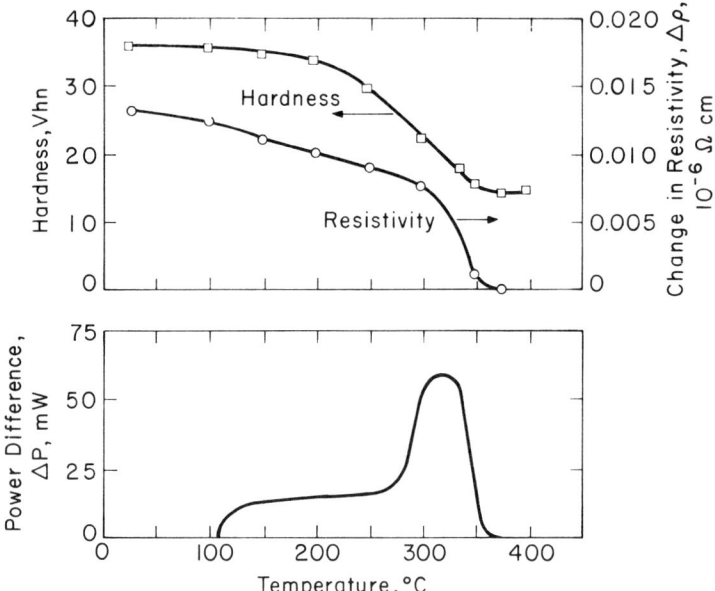

FIG. 57. Rate of energy release, hardness and change in resistivity as functions of annealing temperature for 99.991% aluminum deformed by compression to $\bar{\varepsilon} = -1.39$. Clarebrough et al.[175]

annealing experiments of Vandermeer and Gordon[172] with high-purity aluminum and Åström[105, 106] with 99.99% aluminum. When specimens were given small amounts of slow compression, Mima and Tokizawa observed that only two peaks occurred in the release spectrum; the low-temperature peak evidently had merged with the middle peak. The annealing spectrum of specimens deformed by repeated impact to small strains had four peaks.

The behavior of aluminum differs from that of the noble metals of the same nominal purity (99.99%). Aluminum of higher purity has not been investigated by anisothermal annealing, but Vandermeer and Gordon[172] found three annealing stages above room temperature in the isothermal annealing of aluminum believed by these investigators to approach 99.9999% in purity. The difference between the annealing spectra of aluminum and copper is therefore probably not due to differences in purity.

Åström[105, 106] investigated the isothermal release of energy from 99.99% aluminum deformed by compression. He annealed the same specimen at successively higher temperatures, and obtained a curve of energy release versus time at each temperature. The energy was released in three stages as shown in Fig. 58. The first stage was complete below 100°C, the second extended approximately from 180° to 250°C and the third, which represented recrystallization, occurred at about 350°C. The first stage followed a hyperbolic rate equation; the other two followed exponential equations. The total amount of stored energy measured by Åström appeared to increase with the number of annealing temperatures in his experiments, which is unlikely. Comparison of two sets of results in which eight comparable annealing temperatures were used suggests that the amount of energy released in the first stage was not a function of strain although it occurred at a lower temperature after the larger strain. Increasing the strain from $\bar{\varepsilon} = -0.10$ to $\bar{\varepsilon} = -0.60$ increased the amount of energy released in the second stage by a factor of 10. As with the

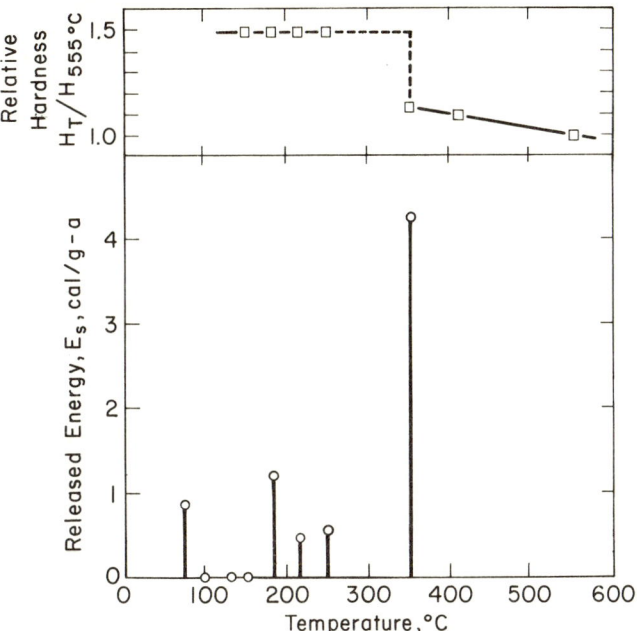

FIG. 58. Released energy and hardness as functions of annealing temperature for 99.99% aluminum deformed by compression to $\bar{\varepsilon} = -0.60$. Annealing carried out on a single specimen at progressively higher temperatures. Åström.[105]

first stage, the temperature at which the energy was released decreased with increasing strain. The increase in strain approximately doubled the energy released during recrystallization. The ratio E_p/E_s increased only slightly from 37 to 42%.

The curves obtained by Sato[73] for aluminum of undeclared purity showed no prominent peak. Evolution of energy began below 50°C and continued to the upper limit of his experiments, namely 550°C, that is, some 200° above the usual recrystallization temperature and only 100° below the melting point of aluminum. Also, the amount of energy reported by Sato was large (up to 100 cal/gram-atom). For these reasons, his results for aluminum are subject to doubt. Possibly they were affected by oxidation during annealing.

Kovacs[83] found a spectrum for the stored energy released by aluminum deformed in torsion to $\gamma_{max} = 0.65$. The spectrum consisted of three stages in the temperature ranges 100° to 150°, 150° to 200° and 250° to 300°C.

The annealing kinetics of 99.99% aluminum between 77°K and room temperature were investigated by van den Beukel[91] (Fig. 49). The spectrum of energy release differed appreciable from the spectra of copper, silver and gold (Fig. 48). Van den Beukel tentatively identified the prominent peak at approximately 210°K with Stage III obtained with the noble metals. He reported poor reproducibility of the amounts of energy released, and suggested that this might be owing to the appreciable recovery at the lowest temperature (about 90°K) at which the calorimeter could be used. Annealing curves reported by Lugscheider and Wildhack[92] for aluminum deformed at 78°K were carried up to 27°C (subsection 4.1.2). They are consistent with van den Beukel's results.

4.2.3.4. *Other Metals*

The stored energy of cold work has been measured by annealing methods in iron (or steel),[73, 77, 93, 97, 143] zinc,[101] cadmium,[87] lead[87] and molybdenum.[190] Except for the work of Sato[73] and Taoka et al.[97] these investigations either gave no information about the kinetics of release or the information related only to part of the release process. Sato[73] investigated vacuum-melted electrolytic iron of unstated purity, Armco iron, and steels containing 0.1 and 0.3% carbon. His results for these, as for other metals, require confirmation before they can be accepted with confidence. In particular, his curves for Armco iron showed large amounts of energy absorbed up to 350°C, an improbable process.

The curves of energy release for two grades of iron cold rolled to 85% reduction ($\bar{\varepsilon} = -1.9$) obtained by Taoka et al.[97] are shown in Fig. 56. Energy was released by the purer iron in a broad peak extending from 100° to 600°C. The peak in the less pure iron extended from approximately 400° to 700°C.

Åström[190] investigated the isothermal release of energy from molybdenum deformed by compression at room temperature. At the three annealing temperatures of 80°, 98°, and 180°C only recovery processes occurred. Except at the lowest strain of 10% ($\bar{\varepsilon} = -0.11$), the release of energy was a hyperbolic function of time. Åström determined from his data for 80° and 98°C an activation energy of 1.5 ± 0.2 eV, which was independent of strain within the limits of accuracy.

4.2.4. *Alloys*

Sato[73] investigated the release of stored energy from a wide variety of nonferrous alloys. These included brass of four compositions, an aluminum bronze containing 7.05% aluminum, a 75Cu–25Ni alloy, a phosphor bronze containing 2.65% tin and 3.89% phosphorus and a 55Cu–26Zn–19Ni alloy. In every instance, the stored energy was released in a complex spectrum having two, three, or four peaks. In many instances the final peak was not the largest.

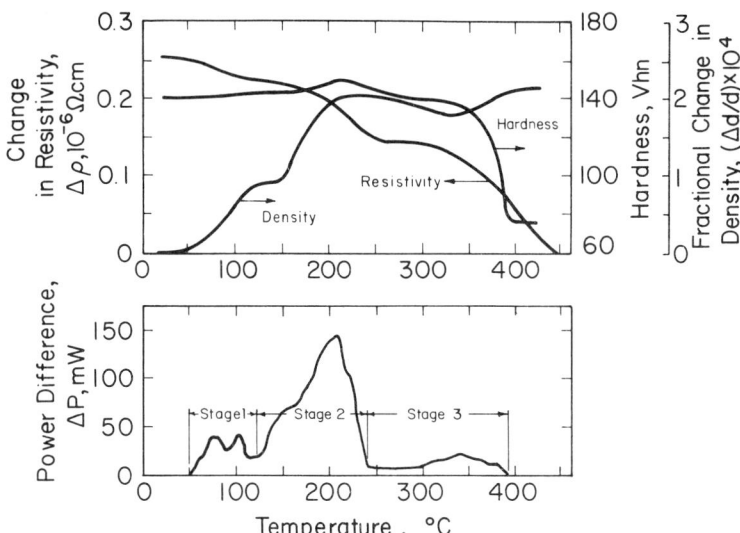

FIG. 59. Rate of energy release, hardness, change in resistivity and change in density as functions of annealing temperature for a 69Cu–31Zn alloy deformed in torsion to $nd/l = 0.47$. Clarebrough et al.[179]

As with nearly pure metals, the final peak, which may be presumed to represent recrystallization, shifted to a lower temperature as the strain was increased. Earlier peaks did not shift appreciably.

Clarebrough et al.[179] investigated the release of energy from a 69Cu–31Zn alloy deformed at room temperature. This was part of an extensive investigation[179, 191–193] into the kinetics of order–disorder phenomena in this alloy. Not the least interesting aspect of this work was that it confirmed very closely Sato's work on a 70Cu–30Zn alloy.

A typical curve of power difference versus temperature obtained by Clarebrough et al. for the 69Cu–31Zn alloy deformed by torsion is shown in Fig. 59. The resolution of the different stages of annealing was better with specimens deformed in tension. Clarebrough et al. divided the annealing spectrum into three stages. The first stage consisted of a double peak between room temperature and 120°C. When the rate of heating was increased from 2°C/min to 6°C/min, the two peaks merged. The amount of energy released in the first stage increased only slightly with increase in strain as shown in Table 4.3. The third stage in the annealing curve resembled the curves for arsenical copper (Fig. 52) and 99.6% nickel (Fig. 45); it consisted of a peak, corresponding to recrystallization, preceded by a long plateau. The amount of energy released in this stage increased with increasing strain. The energy changes in the third stage could not be followed with specimens given small amounts of strain; the recrystallization temperatures were above 450°C and volatilization of zinc caused difficulties in the operation of the calorimeter. Energy was released during the second stage of annealing if the deformation was large, but was absorbed if the deformation was small, as may be seen in Table 4.3.

As already noted, the results of Sato[73] for a 70Cu–30Zn alloy and those of Clarebrough et al. for a 69Cu–31Zn alloy are consistent. This suggests that the results obtained by Sato for various other alloys of copper are likely also to be correct, at least in regard to the pattern of the release of the energy.

Henderson and Koehler[88] measured the release of energy from a 95.5Cu–4.5Zn alloy over the temperature range 113°K to 80°C after compression at 77°K to a strain $\bar{\varepsilon} = -0.58$. The annealing spectrum differed appreciably from that obtained by the same investigators for copper compressed at 77°K to a strain $\bar{\varepsilon} = -0.59$ (subsection 4.1.2). A prominent peak at 263°K in the spectrum of the alloy probably corresponded to the large peak observed at 248°K in that of copper. Details of processes at lower temperatures, corresponding to the multiple peaks observed with copper, were not resolved. The energy release continued over the entire range of temperatures investigated and had not ceased at 80°C. The total amount of energy released by the copper–zinc alloy (21 cal/gram-atom) was over 2.5 times that released by copper strained the same amount.

Van den Beukel[91] investigated the release of energy in the range 77°K to room temperature from alloys of copper with 0.1 and 1.0% silver, 0.1 and 1.0% gold and 0.1% beryllium. Specimens were strained in compression by various amounts at 77°K. The total energy stored in these alloys hardly differed from that stored in copper and appeared to be little affected by the composition of the alloy, but the kinetics of release were altered appreciably. Peaks were less sharply resolved with the alloys than with copper and, according to the identities assigned to them by van den Beukel, were shifted to higher temperatures. The total amount of energy and the amounts allocated by van den Beukel to Stages II and III (see Fig. 48) appeared to increase with strain, but there was appreciable spread.

Averbach et al.[148] investigated the release of energy from deformed gold–silver alloys by isochronal annealing for 1 h at different temperatures. The energy stored in 75Au–25Ag

TABLE 4.3

Energy Released, Change in Electrical Resistivity and Change in Density on Annealing a Deformed 69Cu–31Zn Alloy

Deformation process	Strain, $\bar{\varepsilon}$ or nd/l	Energy release, cal/gram-atom				Decrease in resistivity, $\Delta\rho$, $\mu\Omega$-cm					Fractional increase in density, $(\Delta d/d) \times 10^4$			
		$E_{p(1)}$	$E_{p(2)}$	E_r	E_s	$\Delta\rho_{p(1)}$	$\Delta\rho_{p(2)}$	$\Delta\rho_r$	$\Delta\rho$		$(\Delta d/d)_{p(1)}$	$(\Delta d/d)_{p(2)}$	$(\Delta d/d)_r$	$(\Delta d/d)$
Tension	0.095	4.5	−10.3	—	—	0.014 0.015*	−0.015 −0.005*	0.058 0.053*	0.057 0.063*		0.26	−1.07	−0.91 followed by +0.62	−1.10
	0.19	5.1	1.3	—	—	0.016 0.022*	0.015 0.023*	0.069 0.074*	0.100 0.119*		0.19	−0.69	−0.50 followed by +0.62	−0.38
	0.34	7.1	15	9.6	32	0.040*	0.122*	0.184*	0.346*		0.28	0.63	−0.09 followed by +0.12	0.94
Torsion	0.47 0.47	9.0 7.7	31 39	12 7.7	52 54	0.046* 0.024	0.188* 0.080	0.251* 0.150	0.485* 0.254		0.36 0.78	0.93 1.3	0.57 −0.21 followed by +0.29	1.86 2.16
	1.87	9.0	115	29	153	—	0.408	0.278	—		—	7.5	3.1	—

Subscripts $p(1)$ and $p(2)$ refer to the first and second prerecrystallization stages (see Fig. 59).
* Values obtained with drawn wires.
Ref.: Clarebrough et al.[179]

alloy filings prepared at room temperature was released progressively over the temperature range 150° to about 300°C. Recrystallization took place rapidly above 250°C, while annealing below that temperature appeared to produce only recovery.

Greenfield and Bever[135] investigated the evolution of energy from chips prepared by drilling an 82.6Au–17.4Ag alloy at 78°K and room temperature. The results are shown in Fig. 60. The energy stored in chips prepared at 78°K and annealed at room temperature

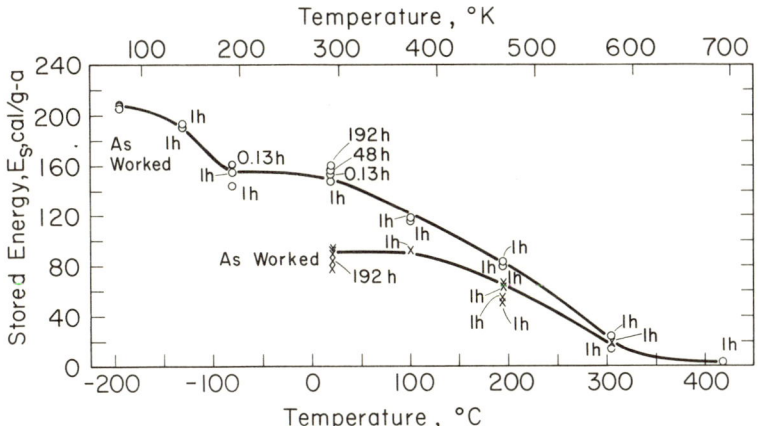

FIG. 60. Stored energy as a function of annealing temperature for an 82.6Au–17.4Ag alloy cold worked at room temperature and at 78°K. Annealing times shown in hours. Greenfield and Bever.[135]

was larger than that stored in chips prepared at room temperature. This has since also been observed in wires of the same alloy[159] (Fig. 23) and gold.[129]

A corresponding relation for 4.2° and 78°K was observed by Appleton and Bever[138] with an 82.6Au–17.4Ag alloy deformed by torsion. Figure 60 shows that the energy was released from the 82.6Au–17.4Ag alloy in two distinct stages. The first of these occurred between 78° and 194°K; the second extended from room temperature up to the completion of recrystallization at about 300°C. This pattern of release may be compared with that of gold shown in Fig. 48. It seems that the evolution of energy that is complete in the alloy at about 195°K can be identified with the first peak of release observed in gold by van den Beukel.[91] It then appears that the process represented by the second peak (Stage III) in gold at about 260°K has shifted to above room temperature in the alloy and overlaps with restoration processes common to specimens deformed at 78°K and at room temperature.

Taoka et al.[97] found two stages of energy release in cold-rolled Fe–2.74Si single crystals and Fe–2.98Si polycrystalline material. In the first stage ranging from 100° to 300°C, the energy released was 53 to 85 cal/gram-atom and was associated with increases in hardness, flow stress and resistivity. In the second stage ranging from 300° to 750°C, the energy released was 125 to 341 cal/gram-atom and was associated with gradual decreases in the hardness, flow stress and resistivity.

Chin and Grant[84, 85] found a single stage of energy release from specimens, prepared by extrusion, of copper containing dispersed particles of alumina.

4.3. VARIABLES RELATED TO THE ANNEALING PROCESS

4.3.1. *Heating Rate in Anisothermal Annealing*

Any thermally activated process occurring while the temperature is increasing continuously takes place over a narrower range of temperature the slower the rate of increase of the temperature. Thus, in anisothermal annealing investigations, sharper resolution of the annealing spectrum is obtained by decreasing the heating rate. This is illustrated by the effect of the reduction in heating rate from 6°C/min to 2°C/min adopted by Clarebrough et al.[179, 191–193] in their investigation of the energy changes in a 69Cu–31Zn alloy, which was mentioned in subsection 4.2.4.

A decrease in the heating rate in anisothermal annealing also shifts any peak to a lower temperature. The magnitude of this shift is a function of the change in heating rate; measurements of the shift can be used to obtain an activation energy, as discussed by Nicholas.[194]

4.3.2. *Temperature in Isothermal and Isochronal Annealing*

The temperature of an isothermal annealing treatment governs the rate at which the restoration processes of recovery and recrystallization proceed. The dependence of the rate of the process on the isothermal annealing temperature forms the basis for standard methods of obtaining activation energies. Every restoration process has a temperature limit below which it does not proceed at a measurable rate. On the other hand, at too high a temperature one process may merge with another. The choice of annealing temperature therefore governs the number of restoration processes that can be observed. This is well illustrated by Åström's investigation of the release of stored energy by aluminum[105, 106] (Fig. 58; see 4.2.3.3).

In isothermal annealing as carried out by Gordon[104] and Åström[105, 106] the kinetics of the restoration processes can be followed at each annealing temperature. In isochronal annealing, the details of the kinetics of each restoration process are lost; only the amount of restoration occurring at any temperature is obtained. Activation energies of particular restoration processes can be calculated from a combination of isothermal and isochronal annealing by means of the analysis proposed by Meechan and Brinkman,[184] which Chuang and Bever[187] applied to the results of an investigation of the restoration of gold at low temperatures.

4.4. ENERGY RELEASE IMMEDIATELY FOLLOWING DEFORMATION

As noted in subsection 3.1.1, Williams,[38, 43, 51–54] using a single-step method with various metals and alloys, observed that measurable quantities of energy were released for an appreciable time after deformation. Figure 61 shows his values[51] of the energy stored in copper compressed by impact. The upper curve gives the amount of energy stored by the specimen 0.01 min after it was struck by the hammers of Williams' apparatus. The lower curve represents the amount of energy stored 1 min after the specimen was struck. The lowest curve in Fig. 62 represents the release of energy in the interval between 0.01 and 1 min after the impact indicated by the two points at the lowest strain in the curves of Fig. 61. The other curves in Fig. 62 represent the release of energy from the specimen after second and succeeding impacts.

Williams[52] also measured the release of energy from lead, aluminum, silver, nickel, iron

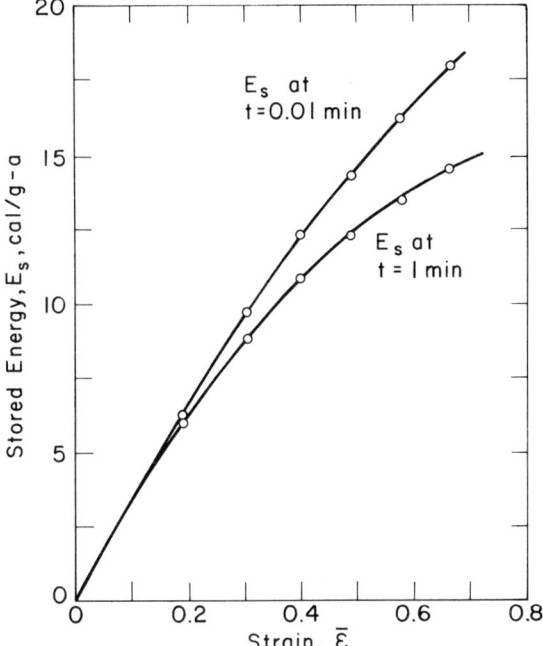

FIG. 61. Stored energy as a function of strain in copper 0.01 and 1 min after deformation by impact. Williams.[51]

and zirconium immediately after deformation. As with copper[38,51] the rates of release decreased progressively with time, except for one specimen of silver. In this specimen, energy was at first released at a progressively decreasing rate, then absorbed briefly and finally released. Table 4.4 summarizes the role of the size of specimen, purity, strain and temperature in the kinetics of the isothermal release of energy observed by Williams for seven metals immediately after deformation.

The release of energy after deformation was also measured by Williams in an aluminum–copper alloy containing 6.7% (14.5 at. %) aluminum and in a 70Cu–30Zn alloy.[53] The release of the energy from both alloys was a function of quenching and aging treatments before deformation. Whereas in nominally pure metals, the rate of release fell to negligible

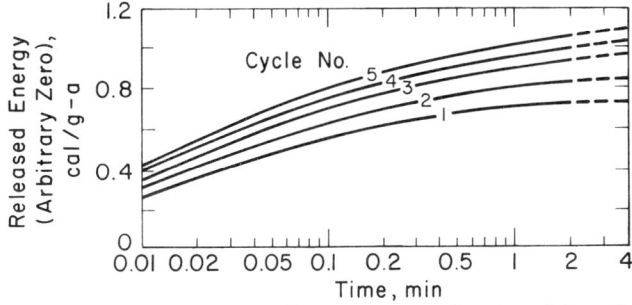

FIG. 62. Amount of stored energy released from copper as a function of time after deformation by impact. The number of the cycle designates the number of blows on the specimen. Williams.[51]

TABLE 4.4

Role of Size of Specimen, Purity, Strain and Temperature in the Kinetics of the Isothermal Release of Energy Immediately Following Deformation

Metal	Size of specimen	Purity	Strain	Temperature
Pb	0	—	0	—
Al	0	0	0	+
Ag	0	—	+	0
Cu	0	0	+	+
Ni	0	0	0	0
Fe	0	0	0	0
Zr	0	—	+	0

Note: 0 = no observed effect.
 + = observed effect.
 − = not investigated.
Ref.: Williams.[52]

values after about 1 min, in the alloys it remained appreciable even after 30 min. The rate of release by brass after certain heat treatments and deformation increased with time during the 30 min for which measurements were made.

Ham[39] followed the temperature changes of copper and Cu–3.8Sn specimens bent plastically by a Charpy pendulum. The temperature of the copper specimen rose sharply at impact and then levelled off to a nearly constant value. After about 130 sec the temperature began to fall. On the other hand, in Cu–3.8Sn the temperature rose sharply at impact but began to fall after about 4 sec. These results suggested that a transient energy release occurred in the copper over a period of 130 sec after deformation but such a release in Cu–3.8Sn is either orders of magnitude slower than in copper or does not occur at all. The first possibility is compatible with the findings of Williams for alloys.[53]

It is a matter of definition whether the energy released as heat immediately after deformation should be regarded as part of the stored energy. The continuing evolution of heat for longer times observed by Williams in alloys poses a different problem. It is also significant that the amount of energy thus released was large (up to 20% of the stored energy).

5. OTHER PROPERTIES MEASURED IN INVESTIGATIONS OF THE STORED ENERGY

The insight that the stored energy can provide into the structural changes due to cold work and annealing is enhanced when other properties are investigated at the same time. Properties that have been measured along with the stored energy include the hardness, flow stress, electrical resistivity, and macroscopic density. In addition, structural changes have been investigated by optical and electron microscopy and X-ray diffraction. Complementary determinations made in investigations of the stored energy are indicated in Table 3.1. The table shows that fewer than one third of all investigations were limited to measurements of the stored energy.

In Part 5, the measurements of hardness, flow stress, resistivity, and density made in conjunction with stored energy measurements are reviewed. The information bearing on the interpretation of the stored energy gained from the investigation of associated structural changes by optical and electron microscopy and X-ray diffraction will be discussed in Part 6.

5.1. Hardness

Measurements of hardness were first combined with measurements of the stored energy by Suzuki in 1949.[14] He observed a sharp decrease in the hardness in the temperature range 270° to 350°C of 99.96% copper compressed to a strain $\bar{\varepsilon} = -0.307$, but reported that at this strain, the observed energy release was complete below 270°C. Suzuki correctly attributed the decrease in hardness to recrystallization. He concluded, however, that all the stored energy was released in recovery processes.

Numerous investigations since Suzuki's have established beyond doubt that recrystallization is associated with a concomitant release of stored energy and a decrease in hardness. This was shown, for example, by the anisothermal annealing investigations of Clarebrough et al. of copper,[46] nickel,[46] silver,[176] gold[176] and aluminum[175]. These investigations also showed that the hardness does not change before recrystallization except when a recovery stage is present in the spectrum of energy release.

The changes in hardness during recovery are generally small, and may be undetectable. For example, with 99.96% copper, in which some recovery took place, Clarebrough et al.[46] found no change in hardness before recrystallization. On the other hand, with 99.55% copper containing 0.35% arsenic, they observed a small decrease in hardness over the extensive recovery range indicated by the release of stored energy (Fig. 52). With 99.6% nickel,[46] the hardness first increased during recovery and then decreased. The increase in hardness was concurrent with the release of energy in a small peak centered at about 250°C (Fig. 45). Bell and Krisement[109] observed no change in hardness during the recovery of 99.9% nickel.

At room temperature in conventional processes of deformation, the effects of cold work on the hardness and the stored energy are roughly parallel. Both quantities increase at decreasing rates as the strain increases. In 82.6Au–17.4Ag and 75Au–25Ag alloys, in which the stored energy levelled off after wire drawing to strains somewhat beyond $\bar{\varepsilon} - 1.0$, the hardness values also levelled off at about the same strains.[128] At all strains the 75Au–25Ag alloy was harder and stored more energy than the 82.6Au–17.4Ag alloy. An initially fine-grained 99.98% copper, which stored more energy than an initially coarse-grained copper of the same batch at all compressive strains up to about $\bar{\varepsilon} = -0.5$, was harder up to about the same strain.[124] Beyond this strain, the initially fine-grained and initially coarse-grained specimens stored approximately the same amount of energy (see subsection 3.2.1) and had approximately the same hardness.

Parallel behavior of the stored energy and hardness was observed in experiments in which the Bauschinger effect or work softening occurred. The decrease in stored energy reported by Iyer and Gordon[157] during secondary compression following primary extension of 99.999% copper was accompanied by a decrease in hardness as shown in Fig. 22. A decrease in hardness also accompanied the decrease in stored energy found by Titchener and Bever[159] during secondary wire drawing at room temperature after primary wire drawing at 78°K (Fig. 23).

With cyclic straining, a parallelism between stored energy and hardness has been found only in some investigations. Clarebrough et al.[161] observed a decrease in the hardness of oxygen-free high-conductivity copper subjected to cyclic straining after a prior reduction by rolling; the stored energy decreased slightly. In an oxygen-free high-conductivity copper subjected to reversed tension, an appreciable decrease in hardness and a release of energy occurred during recovery.[160] No recrystallization was observed, however, and the hardness after recovery was still above the hardness of the annealed material. In a later investigation[161] in the same laboratory, specimens strained cyclically in four-point reversed bending at a stress amplitude of 25,000 psi released energy during recovery and recrystallization with an accompanying decrease in hardness. The decrease in hardness during recovery was more marked in specimens strained cyclically in reversed tension at a stress amplitude of 10,000 psi; in these specimens a second decrease at a much higher temperature (above 500°C) was not accompanied by a detectable release of energy. Recrystallization in the ordinary sense did not occur.

Halford[69] found that the hardness of cyclically strained copper, initially either cold worked or annealed, changed in the same direction as the stored energy. The calorimetric measurements were made by anisothermal annealing. After the energy release the hardness was equal to that of annealed material except in the case of initially annealed specimens in which the hardness after the energy release retained a higher value.

The parallelism between the stored energy and the hardness is not found with explosively loaded specimens. The sharp increase in the energy stored in initially ordered Cu_3Au at shock pressures above 290 kbar, shown in Fig. 20, was not associated with a sharp increase in hardness.[153] The hardness of both the initially ordered and initially disordered alloys increased up to a shock pressure of 160 kbar, but showed only a small increase at higher shock pressures. The initially ordered alloy was harder than the initially disordered alloy except in the undeformed condition.

In the isothermal annealing of shock-loaded 99.999% copper, the stored energy began to decrease before any change in hardness was detected.[151] The contrast between the behavior after shock loading and after extension is shown in Fig. 50.

In an 82.6Au–17.4Ag alloy annealed isochronally after shock loading the hardness decreased in two distinct stages.[152] The larger and sharper of the decreases occurred in the second stage, which was identified metallographically as recrystallization. In contrast, the hardness of filings of a 75Au–25Ag alloy decreased in a single stage corresponding to recrystallization.[148]

An investigation of Clarebrough et al.[179] throws light on the concurrent changes of hardness and stored energy during the anisothermal annealing of a 69Cu–31Zn alloy. The changes in hardness during the annealing of a specimen of this alloy deformed in torsion to a strain $nd/l = 0.47$ are shown in Fig. 59. The hardness increased very slightly during the first stage of the energy release and reached a maximum in the second stage. It decreased somewhat in the latter part of Stage 2 and across the plateau of the third stage. A final sharp decrease accompanied the recrystallization peak of energy release in this stage. Qualitatively similar behavior was found with specimens given a larger torsional strain ($nd/l = 1.87$) and 40 and 60% extension ($\bar{\varepsilon} = 0.34$ and 0.47). In specimens extended 10 and 20% ($\bar{\varepsilon} = 0.09$ and 0.19) the hardness remained constant through Stages 1 and 2, decreased slowly during the early part of Stage 3, and then sharply during recrystallization.

Taoka et al.[97] found in the isothermal annealing of single crystals of the alloy Fe–2.74Si cold rolled 70% ($\bar{\varepsilon} = -1.21$) that the hardness increased slightly in the temperature range of a peak representing the first stage of energy release and decreased appreciably in the range of energy release representing the second stage (recrystallization).

5.2. FLOW STRESS

In many investigations of the stored energy by the single-step method the flow stress or the stress–strain curve has been determined and reported. The earliest investigation in which this was done was by Hort[47, 48] who considered the relation between the stored energy and the stress–strain curve in some detail. His results and those of later investigators are dealt with in section 5.5.

In investigations using a two-step method the flow stress does not have to be measured. When strength properties were considered hardness was usually measured. However, in some investigations the flow stress both of deformed material and of deformed and annealed material was measured as noted in Table 3.1. The relation between the flow stress and the stored energy found in these investigations is also discussed in section 5.5. We need note here only that the flow stress and the stored energy exhibit a generally similar dependence on the extent of deformation.

5.3. RESISTIVITY

The first combined measurements of the stored energy and changes in the resistivity due to cold work were made in 1953 by Clarebrough et al.[189] They found that, during anisothermal annealing, the resistivity of 99.6% nickel deformed in torsion decreased in two stages to the value of annealed specimens. The first decrease was associated with a small peak in the release curve at about 250°C; the second decrease coincided with the recrystallization peak. Similar results obtained subsequently by the same investigators[46] with a similar batch of nickel are shown in Fig. 45.

A sharp decrease in resistivity always occurs during recrystallization. In investigations of the stored energy, this decrease has been observed in various grades of copper[46, 90, 173] and nickel[46, 170] and also in aluminum,[175] silver[176] and gold.[176]

When an appreciable amount of stored energy is released during the recovery of a nominally pure metal, this release is usually accompanied by a small decrease in resistivity. Such a decrease, already noted for 99.6% nickel,[46, 189] has also been observed with 99.96% copper,[46, 173] 99.99% copper[90] and 99.991% aluminum.[175] A large amount of energy was released during the recovery of aluminum. The resistivity of 99.85% nickel decreased in two stages during recovery; each stage corresponded to a peak in the release of energy[170] (Fig. 54). In a batch of 99.6% nickel that showed two recovery peaks the resistivity decreased continuously during recovery; the decrease was sharpest over the temperature ranges of the two peaks.[46] A small release of energy during the recovery of 99.98% copper[174] was not accompanied by any change in resistivity. The release of energy below room temperature from 99.99% copper wires drawn at 77°K was accompanied by a decrease in the resistivity as shown in Fig. 47.[90]

It may be concluded that the pattern of the change of resistivity occurring in a metal during annealing bears a general resemblance to the pattern of change of hardness. The similarity also applies to the effect of the extent of deformation on these quantities. The resistivity increased with strain or expended energy.[66, 87, 90, 129, 195]

Brillhart et al.[154] measured the changes in resistivity during the annealing of shock-loaded copper. They observed three annealing stages. In the first annealing stage the resistivity increased and there was no energy release. In the second stage the resistivity decreased and approximately 10% of the total stored energy was released. In the last stage there was a substantial release of stored energy and a corresponding decrease in resistivity.

The changes in the resistivity during the annealing of a 69Cu–31Zn alloy have been investigated by Clarebrough et al.[179] Annealing after torsional deformation to a strain $nd/l = 0.47$ caused the resistivity to decrease in three stages corresponding to the three stages of energy release shown in Fig. 59. At a larger torsional strain ($nd/l = 1.87$), at which Stages 1 and 2 and the plateau of Stage 3 of the curve of energy release merged, the resistivity decreased continuously throughout the annealing process. In a specimen deformed 20% in tension ($\bar{\varepsilon} = 0.19$) the resistivity decreased in three stages. In specimens extended by greater amounts the stages tended to become less distinct, and after 60% extension ($\bar{\varepsilon} = 0.47$) Stages 1 and 2 merged. Extension to only 10% ($\bar{\varepsilon} = 0.095$) produced a minimum in the resistivity at the end of Stage 1, an increase during Stage 2, a maximum in the region of the plateau of Stage 3 and finally a decrease during recrystallization. Corresponding unusual changes also occurred in the density (section 5.4). The changes in resistivity are included in Table 4.3.

Beardmore and Bever[196] reported a resistivity increase due to torsional deformation of an 82.6Au–17.4Ag alloy at 4.2°K, 78°K, and room temperature. They found a relation between the resistivity increment and the stored energy measured by Appleton and Bever.[138] Beardmore and Bever also investigated the annealing behavior. There were three stages: at temperatures below 220°K the resistivity (measured at 78°K) decreased, between 220°K and 202°C the resistivity increased and at temperatures above 450°C a rapid fall in resistivity corresponded to recrystallization.

Taoka et al.[97] measured resistivity changes over the two stages of energy release in the annealing of Fe–2.74Si single crystals and Fe–2.98Si polycrystalline material. They found a sharp increase of the resistivity over the first stage of energy release and a somewhat less sharp decrease at the beginning of the second stage. Over the second stage the resistivity was either constant or fell slightly. The total effect of the annealing, an increase in resistivity, implies that the final and initial states of the alloys differed. A resistivity increase during a recovery stage has been reported for arsenical copper[46] in which it accompanied a gradual evolution of stored energy.

Cohen and Bever[130] measured the resistivity of initially ordered and initially disordered Cu_3Au as a function of strain at room temperature. The resistivity increased with strain in a manner generally similar to the stored energy, except that the resistivity of the initially disordered alloy passed through a low maximum at low strains. At large strains the resistivities of the initially ordered and initially disordered alloys became approximately equal at room temperature. Their temperature coefficients differed, so that the resistivities at low temperatures were not equal. Roessler and Bever[137] measured the resistivity of initially ordered and initially disordered Cu_3Au as a function of strain at 78°K. The dependence of the resistivity on strain was generally similar to that found by Cohen and Bever[130] after deformation at room temperature. The low maximum at low strains in the initially disordered alloy was absent.

Robinson and Bever[140] found parallel increases in the resistivity and the stored energy of the intermetallic compound AgMg. They observed similar behavior in the compound $TlBi_2$.[183] Gangulee and Bever[180] found parallel increases of the resistivity and stored energy in cold-worked, ordered and disordered Ag_3Mg and silver-rich solid-solution alloys of Ag and Mg. Waldman and Bever[181] made similar observations in silver-rich silver–cadmium alloys. These results will be discussed further in section 5.5.

5.4. DENSITY

In 1934 Maier and Anderson[15] in conjunction with measurements of the energy stored in various deformed materials, measured also the changes in density. They found that hard-drawn wires of copper and aluminum had a lower density than annealed wires. They reported decreases in density expressed as fractions of the densities in the annealed state as 7.8×10^{-4} and 13.3×10^{-4}, respectively. Clarebrough, Hargreaves and co-workers have since published extensive data on the changes in density accompanying changes in the stored energy.[3, 46, 158, 170, 174, 176, 179, 189, 197, 198]

Clarebrough, Hargreaves and co-workers measured the fractional changes in density of copper of four grades,[46, 158, 174] nickel of two grades,[46, 170, 189, 197] silver[176] and gold[176] deformed by various amounts. The density decreased as a result of deformation. The fractional decreases were of the order of 10^{-4} and were larger the larger the extent of deformation.[174, 197] A sharp increase in density always occurred during recrystallization. No other change in the density was observed in those metals in which the only release of stored energy was that associated with recrystallization. In this respect, the changes in density paralleled those in hardness. In metals that released appreciable amounts of energy in one or more recovery stages, the density generally also increased during recovery. Hargreaves et al. observed an exception with 99.99% copper[158] which released approximately half of the stored energy during recovery after deformation by extension to 33% ($\bar{\varepsilon} = 0.29$). Although recovery proceeded continuously above about 80°C, no change in density occurred up to about 220°C.

Figure 54 shows the change in density during the annealing of 99.85% nickel deformed in compression.[170] Similar changes in density were found in 99.6% nickel.[197] In both grades of nickel the density changed continuously over the temperature ranges in which the stored energy was released. In the purer nickel, the density was constant and no energy was released over the temperature range 330° to 400°C.

As has been noted in 3.1.5.1, the stored energy may decrease with increasing strain in the second stage of a two-stage sequence of deformation. Hargreaves et al.[158] observed that the density increased in a small secondary compression of 99.99% copper following a primary extension. This increase of the density accompanied a decrease in the stored energy associated with the Bauschinger effect (Table 4.2).

The relation between the energy and density changes in recovery and recrystallization is complex. In 99.98% copper[174] compressed 55% ($\bar{\varepsilon} = -0.80$) over 20% of the total fractional change in density occurred during recovery; less than 10% of the stored energy was released in this stage. In 99.6% nickel deformed in torsion to $nd/l = 2.34$[46, 197] about 70% of the total change in density occurred during recovery whereas only a little more than 40% of the stored energy was released during recovery. On the other hand, 99.99% copper extended 33% ($\bar{\varepsilon} = 0.29$) regained approximately 27% of the total change of density during recovery, but released 57% of the stored energy.[158]

Brillhart et al.[154] measured the changes in density during the annealing of shock-loaded copper. There were three annealing stages. In the first the density decreased and no energy was released. In the second the density increased and approximately 10% of the total stored energy was released. In the last stage there was a substantial release of stored energy and an increase in density.

The foregoing discussion of density changes has been concerned with nominally pure metals. Clarebrough et al.,[179] in conjunction with measurements of the stored energy, measured the density changes that occurred on deformation of a 69Cu–31Zn alloy. Their results are summarized in Table 4.3. As may be seen, an extension of 10% ($\bar{\varepsilon} = 0.095$) caused an increase in density. (The density changes in Table 4.3 represent restoration processes during annealing of the deformed material, and therefore have opposite signs to those occurring during deformation.) Extension beyond 20% ($\bar{\varepsilon} = 0.19$) produced the usual decrease. Although the overall change in density on annealing specimens deformed by small amounts was a decrease, the usual increase in density occurred during recrystallization. During recovery, the density first increased and then decreased in two stages.

5.5. THE RELATION OF THE STORED ENERGY TO THE FLOW STRESS, RESISTIVITY AND DENSITY

The understanding of the structural changes due to cold work and annealing provided by the measurement of other properties along with the stored energy can be improved by considering the relations between the energy and these properties. The relations can be explored by considering the ratio of one property to another, for example, the ratio $E_s/\Delta\rho$. Alternatively, one property may be considered as a function of the other. Both methods of analysis have been applied to the stored energy, flow stress, resistivity and density.

The broad similarity between the curve of stored energy versus strain and the stress–strain curve has invited comparison of the stored energy with the flow stress from the beginning of investigations of the stored energy. Hort[47,48] showed that the slope of his experimental stress–strain curves, $d\sigma/d\varepsilon$, corrected for "infinitely slow" strain rate, was linearly related to the incremental rate of energy storage $\delta E_s/\delta E_w$. Since $\delta E_s/\delta E_w \approx dE_s/dE_w$, Hort's result is equivalent to the relation $E_s = K(\sigma^2 - \sigma_0^2)$, where K is a proportionality factor and σ_0 the stress at which the stored energy is zero. Fastov[199] proposed an expression of the latter form on theoretical grounds. (See subsection 6.1.2.) Various investigators since have specifically related E_s to the square of the flow stress (Williams,[50,52,54] Bailey and Hirsch,[107] Bailey,[200] Wolfenden,[55,57,59,182] Nakada[68]). When the stress–strain curve is reported (see Table 3.1) published results can be analyzed. In many cases it is found that E_s is proportional to $\bar{\sigma}^2$ and τ^2, where τ is the flow stress in shear. Values of the ratio of E_s to τ^2 for nominally pure metals taken from the results of a number of investigations are given in Table 5.1. For single crystals, τ is the resolved shear stress.

The stored energy was not proportional to the square of the flow stress in all investigations[14] or equivalently dE_s/dE_w versus $d\bar{\sigma}/d\bar{\varepsilon}$ was not a straight line[40,96,119–123]. The stored energy measured by Cohen and Bever[130] for Cu_3Au was not proportional to the square of the flow stress. Wolfenden[182] proposed that the stored energy in Cu_3Au was proportional to the square of the flow stress, but his results did not conform entirely with such a relation. Wolfenden[59] found that the stored energy was proportional to the square of the flow stress in an 82.6Au–17.4Ag alloy.

The relation between the stored energy and the change in resistivity was first considered by Khotkevich et al.[87] They obtained a linear relation between these properties in cadmium and lead deformed at 77°K to different strains and annealed up to room temperature, that is, the ratio $E_s/\Delta\rho$ was constant. Kunin[66] reported constant ratios for silver and copper deformed at room temperature to different strains. Pervakov et al.[195] also found a constant ratio for silver; the value of this ratio was the same for deformation at room temperature and 77°K. More recently Smith and Bever[129] found a constant ratio for gold deformed at 78°K, Gangulee and Bever[180] a constant ratio for three silver-rich silver–magnesium alloys, and Waldman and Bever[181] a constant ratio for four silver-rich silver–cadmium alloys.

Values of the ratio $E_s/\Delta\rho$ computed from the values of the stored energy and resistivity increment reported by various investigators are presented in Table 5.2. This table includes only results of investigations in which $E_s/\Delta\rho$ was constant with strain or for which the variation of the ratio with strain was not reported. The values of the ratio from the investigations of Kunin[66] and Pervakov et al.[195] are much higher than those of Clarebrough and co-workers,[90,170,174–176,179] and Bever and co-workers[129,180,181] because the values of the stored energy obtained by the former were much higher. If these high values of the stored energy are excluded for reasons given in subsection 3.1.1, all values of $E_s/\Delta\rho$ for nominally pure metals lie between 250 and 320 cal/gram-atom-μΩcm except those for 99.85% nickel.[170] The ratio does not depend sensitively on the identity of the metal.

Not all investigators have found $E_s/\Delta\rho$ to be constant with strain. In an investigation of 99.98% copper by Clarebrough et al.,[174] the ratio $E_s/\Delta\rho$ decreased slightly at compressive

TABLE 5.1

The Ratio of Stored Energy to the Square of the Flow Stress in Shear for Nominally Pure Metals

Investigators	Metal	$E_s/\tau^{2\,(a)}$ cal/gram-atom/kg²/mm⁴	$\dfrac{\alpha^2 E_s \mu^{(b)}}{\tau^2}$
Panin and Milevskaya[70]	99.9Cu	0.2	14.2
Clarebrough et al.[124]	99.98Cu	0.04	2.8
Williams[51]	99.999Cu	0.04	2.8
Bailey[200]	99.999+Cu	0.03	2.1
Williams[50]	99.999Cu[c]	0.04	1.8
Williams[54]	99.999Cu	0.05	3.5
Wolfenden[55]	99.999Cu[c]	0.09[d]	4.1
Wolfenden[131]	99.999Cu	0.06	4.2
Bailey and Hirsch[107]	99.99Ag	0.07	2.0
Williams[52]	99.99Ag	0.12	3.5
Nakada[68]	99.99Ag[c]	0.34, 0.24, 0.18[e]	6.3, 4.5, 3.3
Wolfenden[57]	99.92Ag[c]	0.09–0.3[e]	1.7–2.6
Panin and Milevskaya[70]	98.74Ni	0.06	7.9
Williams[52]	99.9Ni	0.016	2.1
Williams[52]	99.998Fe	0.016	1.9
Williams[52]	99.9Zr	0.072	1.9
Nakada[68]	99.99Al[c]	0.4–0.5[e]	6.2–7.8
Smith and Bever[129]	99.99+Au	0.24, 0.28[f]	7, 8.1

[a] τ is taken as $\bar{\sigma}/2$ for polycrystalline metals, and as the resolved shear stress for single crystals.
[b] In this expression $\alpha = 0.5$ for polycrystalline metals and 0.4 for single crystals. E_s is energy/unit volume. See text: 6.3.1.2 (p. 137) for an explanation of the fourth column.
[c] Single crystals.
[d] Average.
[e] Depending on orientation.
[f] At 78°K.

strains beyond $\bar{\varepsilon} = -0.6$ (Fig. 63). Smith and Bever[129] found that, in the deformation at room temperature of 99.99% gold, $E_s/\Delta\rho$ decreased with increasing strain: an average value was about 900 cal/gram-atom-μΩcm. Beardmore and Bever[196] correlated the stored energy measurements on an 82.6Au–17.4Ag alloy by Appleton and Bever[138] with their own resistivity measurements. Deformation was by torsion at 4.2°K, 78°K and room temperature. The ratio $E_s/\Delta\rho$ for deformation at 78° and 4.2°K (followed by a 78°K anneal) decreased during initial straining and attained a constant value of 300 cal/gram-atom-μΩcm at the higher temperature and 260 cal/gram-atom-μΩcm at the lower temperature.

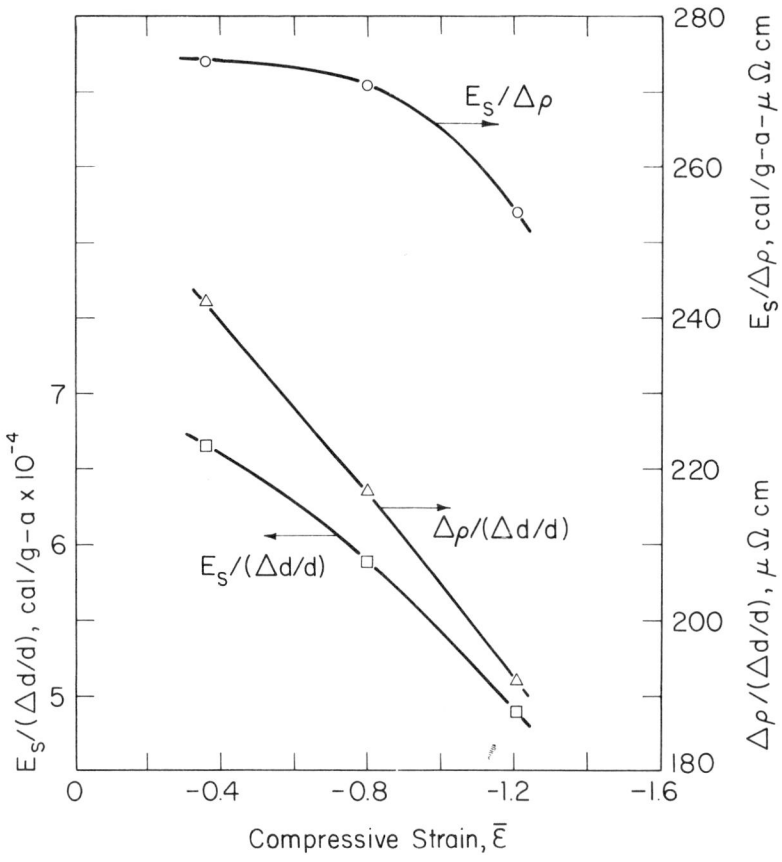

FIG. 63. The ratios $E_s/(\Delta d/d)$, $E_s/\Delta\rho$ and $\Delta\rho/(\Delta d/d)$ as functions of strain in copper deformed by compression. Data of Clarebrough et al.[174]

Robinson and Bever[183] found that $E_s/\Delta\rho$ decreased initially with increasing strain in TlBi$_2$ deformed at 25°C. At large strains, however, the ratio was approximately constant at 8.75 cal/gram-atom-μΩcm. Gangulee and Bever[180] observed a similar trend in a 93Ag–7Mg alloy with long-range order (Ag$_3$Mg). The constant value of $E_s/\Delta\rho$ was approximately 790 cal/gram-atom-μΩcm. In an investigation of the deformation of the intermetallic compound AgMg, Robinson and Bever[140] found that $E_s/\Delta\rho$ increased with

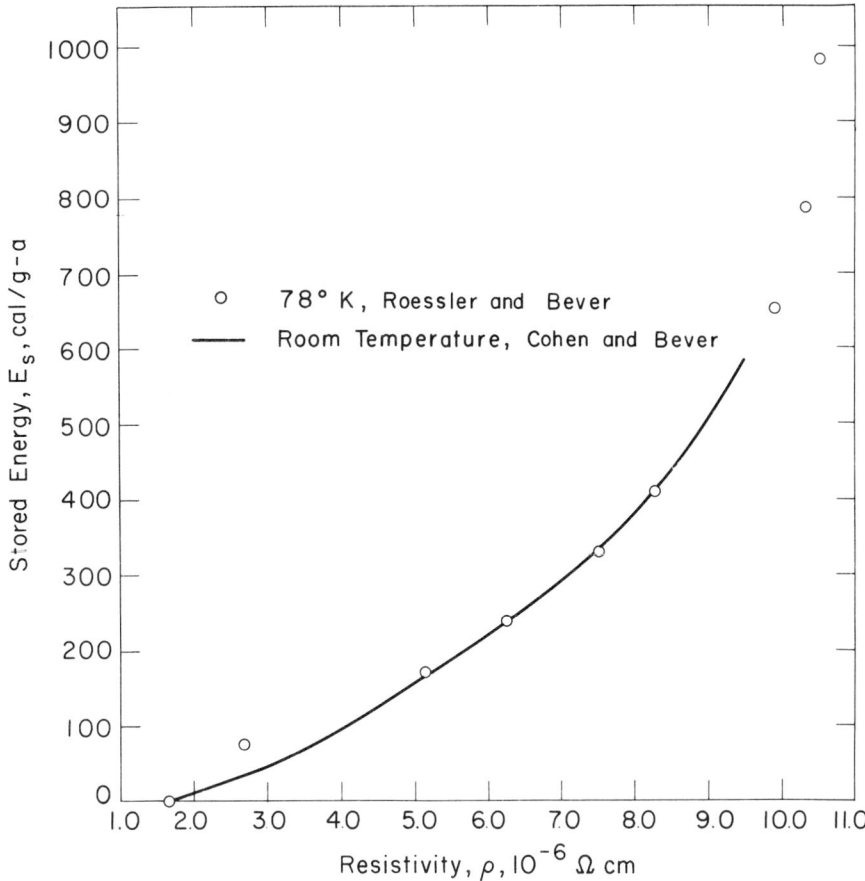

FIG. 64. The stored energy versus the resistivity measured at 78°K of initially ordered alloy Cu_3Au after deformation at room temperature and 78°K by wire drawing to different strains. Cohen and Bever,[130] Roessler and Bever.[137]

strain. The ratio was greater at 25°C than at 165°C and was about the same for a stoichiometric and a nonstoichiometric alloy.

The relation between the stored energy and the resistivity of initially disordered alloy Cu_3Au[130] is shown in Fig. 64. The values for deformation at 78°K and room temperature fall on a single curve. The ratio $E_s/\Delta\rho$ increased continuously with strain. At a given strain, it was larger for deformation at 78°K than for deformation at room temperature.

The ratio of the stored energy to the change in density can be computed for 99.98% copper from the investigation of Clarebrough et al.[174] This ratio is included in Fig. 63. As may be seen, it decreased with increasing strain. Figure 63 also includes the ratio $\Delta\rho/(\Delta d/d)$ which decreased almost linearly with strain over the range investigated.

In 99.6% nickel[46,197] appreciable changes in stored energy and density occurred during recovery, which is in contrast to 99.98% copper. In nickel the ratio of the total stored energy to the total fractional change in density $E_s/(\Delta d/d)$ increased slightly with increasing torsional strain although the ratio for recrystallization $E_r/(\Delta d/d)_r$ appeared to decrease.

TABLE 5.2

The Ratio of the Stored Energy to the Change in Resistivity, $E_s/\Delta\rho$

Investigators	Ref.	Material	Deformation process	True strain, $\bar{\varepsilon}$	Temp of deformation, °K	Ratio, $E_s/\Delta\rho$, cal/gram-atom-μΩcm
Khotkevich et al.	87	Cd	Comp.	—	78	103*
		Pb	Comp.	—	78	300*
Clarebrough et al.	174	99.98Cu	Comp.	−0.36	RT	272
			Comp.	−0.80	RT	270
			Comp.	−1.21	RT	254
V. N. Kunin	66	Ag	Tens.	—	RT	600*
		Cu	Tens.	—	RT	670*
Clarebrough et al.	170	99.85Ni	Comp.	−1.21	RT	93, 118‡
Pervakov et al.	195	Ag	Lat. comp.	to 0.7†	RT	850*
			Lat. comp.	to 0.5†	77	850*
Clarebrough et al.	175	99.991Al	Comp.	−1.39	RT	246
Clarebrough et al.	176	99.98Ag	Comp.	−1.39	RT	306
		99.9Au	Comp.	−1.39	RT	281
Loretto et al.	90	99.99Cu	Wire dwg.	0.44	RT	270
				0.44	77	318, 284‡
Clarebrough et al.	179	69Cu–31Zn	Tens.	0.92	RT	109, 49‡
Smith and Bever	129	99.99Au	Wire dwg.	to 1.8	78	270*
Gangulee and Bever	180	Ag–13Mg	Wire dwg.	to 1.0	RT	250*
		Ag–5.4Mg	Wire dwg.	to 1.0	RT	250*
		Ag–7Mg (disordered)	Wire dwg.	to 1.5	RT	250*
Waldman and Bever	181	91Ag–9Cd	Wire dwg.	2.6	RT	120*
		75Ag–25Cd	Wire dwg.	2.6	RT	80*
		67Ag–33Cd	Wire dwg.	2.6	RT	75*
		61Ag–39Cd	Wrie dwg.	2.6	RT	85*

* Ratio constant with strain.

† Fractional decrease in diameter of wire deformed by lateral compression.

‡ First value is for total stored energy and change in resistivity; second value is for recrystallization only.

Chuang and Bever[187] measured the ratio of the energy released during an annealing stage to the change in resistivity, $\Delta E_s/\Delta\rho$, in gold deformed by wire drawing at 78°K. The ratio in Stage II annealing (78°K to 195°K) was 263 cal/gram-atom-μΩcm, in Stage III (195°K to 300°C) 186 cal/gram-atom-μΩcm, and in Stage V 263 cal/gram-atom-μΩcm.

6. INTERPRETATION

A successful theory of the stored energy of cold work must be able to explain the mechanisms by which energy is stored and released and must account for the effects of variables on the amount of the stored energy and on the kinetics of its release. The literature contains various theories of the stored energy. While some of these are no longer tenable, others still contribute to the interpretation of the stored energy of cold work.

In section 6.1 we shall review the development of the interpretation of the stored energy from a historical viewpoint. Section 6.2 is a brief summary of the structural characteristics of cold-worked metals. In section 6.3 we shall review systematically current interpretations of the stored energy. Section 6.4 is concerned with the relation of the stored energy to the kinetics of recovery and recrystallization, section 6.5 deals with the mechanisms by which energy expended in deformation is dissipated as heat and section 6.6 summarizes the contributions of different types of imperfections to the stored energy.

6.1. Historical Survey

Hirn,[201] in experiments carried out about 1855, determined the mechanical equivalent of heat by equating the work expended and the heat evolved in the compression of lead. His determination was criticized on the grounds that the deformation altered the state of the metal. His critics, however, did not advance definite ideas to explain what this alteration was and Hirn argued that in lead it was insignificant. This controversy appears to be the earliest reported instance in which possible changes in the internal energy of a metal due to cold working were considered.

In reporting the first measured values of the stored energy of cold work, Charbonnier and Galy-Aché[49] did no more than attribute the change in internal energy to the effects of cold work. The first attempts at an interpretation of the stored energy date from the work of Hort.[47, 48]

Four concepts have provided the framework on which all theories of the stored energy have been built:

(1) the formation of a new "phase" during cold work;
(2) the existence of a functional relation between the stored energy and some other property;
(3) the introduction of lattice strains;
(4) the creation of lattice imperfections.

The last three of these concepts need not be mutually exclusive in the interpretation of the stored energy of cold work.

6.1.1. Interpretation Based on a Phase Change

Hort[47, 48] tried to explain the stored energy as a heat of transformation. He drew support for this explanation by citing as comparable the heats of certain solid-state transformations the existence of which has since been disproved. The idea of a phase change remained alive, however, during the currency of the "amorphous metal" theory. Rosenhain, a strong advocate of this theory, wrote in 1923[202] that "the theory that there is a phase change (crystalline to amorphous) which occurs under plastic strain has recently received interesting confirmation, although only of a preliminary sort, from some experiments carried out by Dr. Sinnatt". As noted in Table 3.1, however, the results of these experiments are of doubtful validity.

Russell[203] sought to explain the difference between the amount of work actually required to cause plastic deformation of iron and that necessary to cause the same amount of deformation in the absence of work hardening as absorbed in creating a "vitreous phase". He deduced a value of the stored energy on the assumption that the latent heat of fusion of this vitreous phase, assumed to be equal to the heat of fusion of iron, was retained. It is now well established that the amount of energy stored during cold work is considerably smaller than the amount of work expended in producing work hardening.

Maier and Anderson[15] postulated the formation of a new "ω-phase" to explain parasitic electromotive forces observed in a thermocouple under stress. The amount of this phase computed independently from their values of the stored energy and the change in density, however, did not agree. A later investigation by Maier[204] of changes of density with cold work led him to the conclusion, contrary to that of Maier and Anderson, that the "ω-phase" was denser than the annealed metal.

Under certain conditions and in specific metals or alloys, stress may induce phase transformations of a martensitic type and may accelerate other transformations. However, phase transformations are in general not a result of cold working, and the concept of an amorphous phase produced by cold work has lost all standing. The interpretation of the stored energy of cold work in terms of a phase change, therefore, retains only historical interest.

6.1.2. Interpretation Based on Functional Relations

As discussed in section 5.5, the broad similarity between the curve of stored energy versus strain and the stress–strain curve has invited correlations of the stored energy with the flow stress from the beginning of investigations of the stored energy. Hort[47, 48] showed that the slope of the stress–strain curve $d\sigma/d\varepsilon$ for an "infinitely slow" extension was linearly related to the incremental rate of energy storage $\delta E_s/\delta E_w$. Since this rate approximates the instantaneous rate found by Hort, dE_s/dE_w, the linear relation is equivalent to the relation

$$E_s = K(\sigma^2 - \sigma_0^2) \tag{6.1}$$

where K is a proportionality factor and σ_0 represents the stress at which the stored energy is zero. Fastov[199] proposed an expression of this form on theoretical grounds. From his assumptions, he deduced that $K = \lambda/2Y$, where λ was a constant of the material and Y Young's modulus. Fastov held σ_0 to be identical with the yield stress, σ_y. Relations of similar form were obtained by various investigators.[50, 52, 54, 55, 57, 59, 68, 107, 146, 181, 182, 200] These findings and their implications will be discussed in subsection 6.3.1.

Taylor and Quinney,[40] like Hort, correlated the incremental rate of energy storage $\delta E_s/\delta E_w$ with the slope of the stress–strain curve. Their values of $\delta E_s/\delta E_w$ were almost constant up to large strains, and then fell rapidly towards zero. These findings are contrary to those of Hort, Williams, and others quoted above, but agree fairly closely with those of Kunin and Senilov,[119] Fedorov[120] and several other investigators.[14, 96, 120–123]

McAdam[205, 206] analyzed the values of the stored energy of Taylor and co-workers[40, 77, 169] and Rosenhain and Stott[41] in relation to his theory of intrinsic strength.[207] According to this theory, the "state of strength" of a metal could be defined solely by the stress required to cause it to flow, regardless of the prior history of strain and temperature. On the assumption that the stored energy was a function only of the state of strength, McAdam deduced for copper the general form of the curves of E_s and dE_s/dE_w as functions of strain. His theory, however, implies the existence of a mechanical equation of state for a metal, which is not generally valid.

Fedorov[120] and Studenok[122] explained the stored energy as arising from the simultaneous actions of hardening and relaxation. Energy storage accompanied the former, while the latter released energy. The relation of the stored energy to the melting point,[120] the higher value of E_s/E_w for dynamic than for static compression,[122] and the smaller amount of energy stored after prior deformation[122, 123] were all qualitatively explained on this basis.

Several authors attempted to calculate the stored energy from a hypothetical temperature rise associated with the measured change of a property. For example, Geiss and van Liempt[208, 209] and later van Liempt[210] computed the temperature increase that would be necessary to bring about an increase in electrical resistivity equal to that caused by a given amount of cold work. The stored energy was then obtained as the product of this hypothetical temperature rise and the heat capacity. Wood[211] made a similar computation using the hypothetical temperature rise required to cause a lattice expansion equal to that caused by cold work and measured by X-ray diffraction. Boas[212, 213] equated the effects of temperature and cold work on the intensity of X-ray reflections to obtain values of the stored energy. Taylor[213a] pointed out that the energy due to the elastic strains determined by Wood from his X-ray data was much smaller than the stored energy of cold work. He concluded that most of the stored energy of cold work must be concentrated in regions very close to the dislocations.

Interpretations in which analogies were postulated between the effects of temperature and cold work on some selected property now have only historical interest. The responses of a crystal lattice to cold work and to an increase in temperature are basically different and the notion that the cold-worked state could be described by such a concept as "frozen heat motion" is now seen to lack any physical basis.

The direct comparison of changes in the stored energy and changes in some other property such as the flow stress (section 5.2) or the electrical resistivity (section 5.3) may make possible the prediction of the strain dependence of the stored energy. Faulkner[214] demonstrated that in 99.98% copper deformed by rolling the increase in the mean-square width of nuclear magnetic resonance lines appeared to be proportional to the stored energy. Faulkner and Ham[215] assumed that this proportionality also held for 99.991% aluminum deformed by rolling. They estimated the energy stored in aluminum at various strains from the further assumption that the energy stored after rolling to a reduction in area of 88% ($\bar{\varepsilon} = -2.13$) would be approximately equal to that stored after 75% compression ($\bar{\varepsilon} = -1.39$).[175]

6.1.3. Interpretation Based on Lattice Strains

X-ray diffraction techniques have played an important part in the investigation of the structure of cold-worked metals. In the 1930's papers appeared[216–219] in which X-ray data were used in the calculation of the elastic strain energy resulting from lattice distortion caused by cold work. During this period doubt arose as to the interpretation of X-ray line broadening due to cold work. In particular, Wood[220] attributed broadening to lattice fragmentation. Smith and Stickley[221] and Stokes, Pascoe and Lipson[222] attempted to decide between the lattice distortion and the fragmentation theories. Although their results showed appreciable scatter, Smith and Stickley and Stokes et al. concluded that the main cause of broadening was lattice distortion.

The problem was investigated again by Warren and Averbach[223] who took advantage of the sensitivity of X-ray spectrometer techniques as compared with film techniques. They analyzed the effect of cold work on X-ray patterns using Fourier analysis and developed a method of separating lattice distortion and particle size effects.[224–227]

It has been assumed[148] that the value of particle size obtained from the analysis of line broadening represents the size of subgrains and that the subgrains or particles are separated by low-angle boundaries consisting of dislocation arrays. A difficulty arises because the values of particle size and root-mean-square strain depend on the assumed distribution of strain.[149] A more serious difficulty is that the cells revealed by transmission electron microscopy of thin films of cold-worked metals are about an order of magnitude larger than the particles of the X-ray analysis. This problem will be discussed in 6.3.1.4.

It has also been assumed[149] that the stored energy computed from the root-mean-square strain obtained from X-ray line broadening should be equal to that part of the stored energy which is due to dislocations. This will be discussed in 6.3.1.5.

6.1.4. *Interpretation in Terms of Lattice Imperfections*

The application, in 1934, of the concept of dislocations to the deformation of metals was soon followed by the interpretation of property changes in terms of imperfections. Boas,[212] although not using the term "dislocation", proposed that the bulk of the stored energy must reside in small inhomogeneously deformed regions rather than in the elastically distorted, but homogeneous lattice. Burgers[228] also discussed such a model; he postulated coherent but elastically distorted lattice blocks containing most of the atoms and, between these blocks, inhomogeneous distorted transition regions, which stabilized the residual stresses existing in the material. He concluded from available X-ray evidence[212, 216, 218, 219] that only a small fraction of the total stored energy could be identified with the elastic energy of the homogeneous strained blocks, and that the balance of the stored energy must be associated with the inhomogeneous distorted regions surrounding them.

Koehler[229, 230] and Seitz and Read[231, 231a] were the first to apply the dislocation theory to the stored energy of cold work. Seitz and Read took the energy of a dislocation, in analogy with that of an interstitial or a vacancy, to be of the order of 1 eV per atom plane pierced. From the experimental values of the stored energy obtained by Taylor and Quinney[40, 77] they deduced that, in a heavily deformed metal, about 1 atom in every thousand was near a dislocation center. This is equivalent to a dislocation density of about 10^{12} lines/cm^2. They calculated a similar value from Taylor's theory of work hardening. Koehler[229] computed from an analysis of the elastic strain field, that the energy required to create a dislocation was several electron-volts per atom plane pierced. By equating the strain energies of various arrays of dislocations with the stored energy values of Taylor and Quinney,[40] he deduced that the average spacing between dislocations was of the order of 10^{-6} cm, that is, the dislocation density was about 10^{12} lines/cm^2.

Bragg[232] held that the stored energy of cold work could not be accounted for by elastic strain energy, but that it was probably due to the creation of interfaces, consisting of a series of dislocations, between mosaic blocks or subgrains. He proposed a two-dimensional model and estimated that, if a metal were to store energy equal to the amount measured by Taylor and Quinney,[40] interfaces must have been produced in which dislocations were spaced as closely as possible, that is, about 1 dislocation in every 2.5 atom planes. Because of the model he adopted, Bragg avoided assumptions about the size of the mosaic blocks. If the size of a block is of the order of 10^{-5} cm, the energy of the dislocation per atom plane pierced becomes 5–6 eV and for an array of dislocations such as Koehler assumed, the dislocation density necessary to produce an amount of stored energy equal to that measured by Taylor and Quinney[40] is approximately 10^{12} lines/cm^2, in agreement with the estimates of Seitz and Read[231, 231a] and Koehler.[229]

The early models of the cold-worked structure of a metal, involving Taylor or similar arrays of dislocations, are now known to be inadequate. Also, other crystal imperfections may contribute to the energy stored during cold working. Nevertheless, the analyses of Seitz and Read,[231, 231a] Koehler[229] and Bragg[232] focused attention on the relation of dislocation density to the stored energy and served as a starting point for subsequent attempts at interpreting the stored energy in terms of dislocations.

6.2. The Relation between Variables and the Structure of Cold-worked Metals

An adequate understanding of the structure of cold-worked metals began with the emergence of dislocation theory. The current state of knowledge, which is the outcome of almost 40 years of speculation, analysis and experiment, gives a detailed qualitative representation of the cold-worked state, but is still incomplete in its quantitative aspects. Much of the experimental evidence has emerged as a result of the application, beginning in 1956,[232a, 233] of transmission electron microscopy to thin films of metal. Until then most of the evidence had been drawn from the interpretation of etch-pitting experiments and X-ray diffraction data. The X-ray data often led to conflicting conclusions. Transmission electron microscopy has largely resolved these conflicts and, in particular, has shown that the structure of a cold-worked metal as deduced from X-ray microbeam investigations[234–237] was basically correct.

In this section we shall briefly describe those features of the structure of cold-worked metals that are of special relevance to the interpretation of the stored energy of cold work. The section will be based on investigations and reviews that do not primarily concern themselves with stored energy. No complete description of the cold-worked structure nor an exhaustive review of the available literature is intended here. Reference is made to publications by Hirsch,[238] Swann,[239] Keh and Weissmann,[240] Nabarro, Basinski and Holt,[241] Feltner and Laird[242] and Steeds.[243]

Since the structure is affected by the same variables that affect the amount of the stored energy and the kinetics of its release, this discussion is organized under the same headings as those used in Parts 3 and 4.

6.2.1. Extent of Deformation

In the first 10 to 25% strain the dislocations in f.c.c. and b.c.c. metals not only multiply but also tend to arrange themselves in clusters. When a foil of a worked metal is examined by transmission electron microscopy the dislocations are found to be gathered together in dense, irregular tangles that are separated by relatively dislocation-free regions. At the end of the first 10 to 25% strain these dense irregular tangles have joined to form continuous walls—cell walls—which separate the cells. The number of dislocations in a cell wall is greater than the number in a simple tilt or twist boundary.

With increasing strain the average size of the cells decreases. The rate of decrease of the cell size with strain decreases as the strain increases.[244, 245] In copper the cell size has been observed to decrease by a factor of approximately 2 between a strain of $\bar{\varepsilon} = 1$ and $\bar{\varepsilon} = 4$; in iron over the same strain interval the size decreases by a factor of about 5.[244] In f.c.c. metals the cell size appears to have a greater tendency to level off with increasing strain than in b.c.c. metals. This leveling off is associated with the coalescence of the cells. Cell formation is not usually observed in cold worked h.c.p. metals.

Dislocation densities in metals cold worked to strains of less than 25% have been determined by electron microscopy. The dislocation density has been found to increase with strain but at a rate that decreases with strain.[240, 246] Evidence suggests that the dislocation density and the cell diameter are inversely related.[247, 248]

The extensive investigations by electron microscopy of the structure of cold-worked single crystals, particularly after deformation in Stage II of the stress–strain curve will be discussed in section 6.2.8.

6.2.1.1. *Recovery*

According to Keh and Weissmann[240] the cell walls in deformed iron during recovery tend to become regular dislocation networks. Simultaneously, the average density of dislocations decreases. These findings contrast with those of Bailey and Hirsch,[107] who observed that the dislocation density did not change perceptibly during the recovery of silver. The metals investigated and the temperature of annealing, however, were different.

6.2.2. *Temperature of Deformation*

A dislocation cell structure is produced by deformation at low temperatures just as it is at room temperature. The cell walls are less clearly defined after deformation at low temperatures. At a given strain the cell diameter is smaller and the dislocation density larger after deformation at a low temperature than after deformation at room temperature. The rate of work hardening is greater at low temperatures.

6.2.3. *Rate of Deformation*

Edington[249, 250] examined the dislocation structure of niobium and copper single crystals deformed to maximum strains of 30% at rates between 10^{-4} and 10^4 $\sec^{-1}$. The dislocation density at the same strain was unaffected by the strain rate.

6.2.4. *The Deformation Process*

All deformation processes which are carried out at conventional strain rates produce a dislocation cell structure in the deformed metal, with the possible exception of h.c.p. metals. The detailed features of the structure, particularly the misorientation across the cell walls and hence their energy, depend on the deformation process, but the nature of this dependence is not yet generally known. Large gradients of plastic strain as in orthogonal cutting or in the wire-drawing of b.c.c. metals in which the grains become ribbon-shaped and curly[251] would of geometric necessity cause a large misorientation across the cell walls. In processes such as extension or compression, on the other hand, the plastic strains are more uniform and the misorientation should be smaller. On a macroscopic level the deformation process affects the cold-worked structure by controlling both the development of crystallographic texture and deformation banding.

The structure of shock-loaded metals and alloys has been investigated by microscopy and X-ray diffraction.[154, 252–256] In shock-loaded f.c.c. metals dislocations form a cell structure which appears to be less well defined than the cell structure produced by conventional deformation. Twinning is also observed; the density of twins in general increases with the shock pressure. In b.c.c. metals the dislocations are mainly long and parallel screw dislocations.

Cohen, Nelson and De Angelis[257] believed that the twins in shock-loaded copper had incoherent interfaces. They observed that recrystallization began at twin intersections and continued along twin interfaces. They found no change in the microstructure before recrystallization.

6.2.5. Deformation History

6.2.5.1. Bauschinger Effect

The effect of reversed deformation on the dislocation structure does not appear to have been investigated.

6.2.5.2. Work Softening

Keh and Weissmann[240] reported that the fairly uniform dislocation structure in iron deformed 13% at 138°K was replaced by a cell structure after a subsequent 19% strain at room temperature. The cell structure appeared to be the same as if the specimen had been deformed to the same total strain at room temperature.

6.2.6. Cyclic Deformation (Fatigue)

The dislocation structure produced by cyclic straining at "small" strain amplitudes differs from that produced at "large" strain amplitudes. Feltner and Laird[242] define "small" strain amplitudes as those for which the fatigue life is greater than 10^6 cycles and "large" strain amplitudes as those for which the fatigue life is smaller than 5×10^4 cycles. At small amplitudes the structure consists almost entirely of dislocation dipoles. The density of dipoles increases and the dipoles tend to cluster as the cumulative strain increases. A cell structure is not observed. At large amplitudes a cell structure forms and and appears to be similar to that after noncyclic deformation. Some of the boundaries, however, are mainly made up of dipoles especially at small strain amplitudes.

6.2.7. Grain Structure

At small strains the dislocation density in fine-grained metals has been observed to be greater than that in coarse-grained metals deformed to the same strain.[240, 246, 258, 259] Edington and Smallman[259] found that in vanadium deformed in tension to a strain of 3% the square root of the dislocation density was proportional to the inverse square root of the grain size.

6.2.7.1. Single Crystals

The dislocation structure in f.c.c. single crystals strained into Stage I consists of clusters of short primary edge dislocation dipoles. These clusters are sometimes called multipoles or braids. The density of secondary dislocations is approximately one-tenth that of primary dislocations.[243, 260, 261] The dislocation density N increases linearly with shear strain γ in easy glide. $dN/d\gamma$ is approximately 10^8 lines/cm^2. At the end of easy glide, the whole crystal is filled with primary dipoles and multipoles.

The transition to Stage II is marked by widespread secondary slip. The density of secondary dislocations becomes approximately equal to the density of primary dislocations. A cell structure forms: the cells are elongated along the primary plane and their dimensions are consistent with the slip distances observed on the surface of the crystal. The average cell size decreases with increasing strain. The cell walls are believed to be the result of dislocation pile-ups and secondary slip arising from the stress concentration at the pile-up. Tilt and twist boundaries and multipole walls result. Thus the crystal consists of "hard" and "soft" regions.

In Stage III the boundaries of misoriented regions become more sharply defined and the misorientation increases. Fine surface slip lines seen in Stage II are replaced by well-defined coarse slip bands. In Stage II and Stage III the shear stress is given by $\tau = \alpha \mu b N^{\frac{1}{2}}$

where α is approximately 0.4 if primary and secondary dislocations are considered separately and approximately 0.3 if the total dislocation density is considered.[262]

6.2.8. *Purity*

There appears to have been no systematic study made of the influence of purity on the dislocation structure of cold-worked metals.

6.2.9. *The Identity of the Metal*

A dislocation cell structure forms in both f.c.c. and b.c.c. metals. A cell structure is seldom observed in h.c.p. metals. Differences in the dislocation structure of metals of the same crystal structure are attributable to differences in melting point and stacking fault energy. The slip systems in f.c.c. and b.c.c. metals are different. This causes a difference in the crystallographic texture that develops during deformation which in turn may affect the dislocation structure.

6.2.10. *Alloys and Intermetallic Compounds*

The stacking fault energy may be lowered by the addition of solute. The separation of partial dislocations becomes larger. In alloys of low stacking fault energy (*ca.* 5 erg/cm^2), dislocations tend, at small strains, to be in planar arrays rather than in tangles.

The dislocation density at a given strain is larger in alloys than in nominally pure metals and the cell size is correspondingly smaller.

6.2.10.1. *Alloys with Long-range Order*

Superlattice dislocations are a characteristic feature of deformed long-range ordered alloys. The antiphase domain boundary probability increases with increasing strain.

6.3. Interpretation of the Effects of Variables on the Stored Energy

In this section we shall consider the interpretation of the stored energy in terms of the effects of variables. The variables will be taken up in the same order as in Parts 3 and 4.

6.3.1. *Extent of Deformation*

The flow stress of cold-worked metals is closely related to the dislocation structure. In nominally pure metals, the increase of E_s and decrease of dE_s/dE_w with expended energy so closely parallel the increase of $\bar{\sigma}$ and decrease of $d\bar{\sigma}/d\bar{\varepsilon}$ with strain that the stored energy must also be strongly related to this dislocation structure. In metals of high purity it has been repeatedly observed (subsection 4.1.1) that all or nearly all of the stored energy is released during recrystallization. The long-standing assumption that recrystallization is attended by the removal of a large excess of dislocations introduced by cold working has been confirmed by transmission electron microscopy.[263, 264] The stored energy released during recrystallization, therefore, represents dislocations which are eliminated during the restoration of the annealed structure. This is undoubtedly as true for impure metals and alloys as it is for metals of high purity, although several processes may be going on simultaneously in the former (subsection 6.3.10).

Theories of the increase in stored energy with strain are of two kinds. Those of the first kind relate the stored energy to the dislocation structure produced by cold work and consider the manner in which this structure is developed. (Contributions by point defects to the stored energy will be discussed later.) Theories of the second kind relate the stored energy to the dislocation structure but ignore the process of formation of this structure.

6.3.1.1. Stored Energy and Strain

Theories of the first kind depend on specific models of the dislocation structure and of the mechanism by which the structure develops. In general the theories were conceived primarily as theories of work hardening, that is as explanations of the dependence of the flow stress on strain. Moreover, some are concerned only with the work hardening of single crystals in Stage II of the stress–strain curve; these latter theories will be discussed in 6.3.7.1.

Koehler[229] computed the elastic strain energy of dislocations arranged regularly on a two-dimensional "Taylor lattice". He also computed the elastic strain energy of dislocations arranged in pairs or dipoles.[230] For both arrangements the elastic strain energy was proportional to the dislocation density. Following Taylor[265] he assumed that the density of dislocations increased linearly with strain. As a consequence he predicted for Taylor's parabolic work hardening ($\bar{\sigma} \propto \bar{\varepsilon}^{\frac{1}{2}}$) that the stored energy should increase approximately linearly with strain, but this is not generally found.

Seitz and Read[231] assumed that the energy of a dislocation was approximately equal to that of an interstitial or a vacancy and hence of the order of 1 eV per atom plane pierced. They predicted as did Koehler, that under conditions of Taylor hardening the stored energy due to the elastic strain energy of dislocations should increase linearly with the strain.

Stroh[266] computed the elastic strain energy of groups of piled-up dislocations. He found that the strain energy was proportional to the number of groups and approximately proportional to the square of the number of dislocations in each group. Since he followed a theory due to Mott[267] which predicted parabolic work hardening, he obtained a linear relation between the elastic strain energy and the plastic strain, similar to the result obtained by Koehler and by Seitz and Read. From his analysis and experimental stress–strain curves Stroh deduced that the incremental rate of energy storage $\delta E_s/\delta E_w$ was of the order of 5%.

Most of the subsequent theories relate to Stage II of single crystal deformation and will be reviewed in 6.3.7.1. Kuhlmann-Wilsdorf[268] has considered the deformation of single crystals in Stage III and Langford and Cohen[245] have considered the deformation of polycrystalline metals. Kuhlmann-Wilsdorf has also analyzed the deformation of iron at large strains.[269]

Kuhlmann-Wilsdorf expressed the elastic strain energy of dislocations arranged in cell walls as

$$E \approx N \frac{\mu b^2}{4\pi} \ln \frac{N^{-\frac{1}{2}}}{b} \qquad (6.2)$$

which is similar to eq. (1.17) in which the outer cut-off radius $R = N^{-\frac{1}{2}}$ and the core radius $r_0 = b$ and where N is the dislocation density, μ the shear modulus and b the Burgers vector. Kuhlmann-Wilsdorf's mesh-length theory for Stage III single-crystal deformation which relates the flow stress to the stress required to bow dislocations out between pinning points against their line tension predicts a linear increase in the dislocation density with strain. Accordingly, the stored energy would increase approximately linearly with strain. A flow stress proportional to $N^{\frac{1}{2}}$ and a parabolic stress–strain curve are also predicted.

Thus the conclusion of this theory is the same as that of the theories of Koehler, Seitz and Read, and Stroh. Like Stroh's theory, Kuhlmann-Wilsdorf's predicts from experimental stress–strain curves that the value of E_s/E_w is small. The theory also has the attraction of being based on a realistic model of the dislocation structure.

Kuhlmann-Wilsdorf[269] applied the mesh-length theory, in the form appropriate to Stage II deformation of single crystals, to the results of Langford and Cohen for drawn iron wire.[245] These investigators had found linear hardening up to strains of $\bar{\varepsilon} = 7$. Kuhlmann-Wilsdorf's theory predicts that the stored energy E_s is a small fraction of the expended energy and increases as the square of the strain. Such a concave-upward curve would be most unusual.

Langford and Cohen[245] suggested that the cell wall energy in drawn iron was about 500 erg/cm². The energy of the cell walls per unit volume is $(\pi/2\bar{d})$ 500, where $\bar{d}$ is the cell diameter. From the measured value of $\bar{d}$ they estimated that the stored energy in wire drawn to $\bar{\varepsilon} = 6$ would amount to about 0.2% of the expended energy. Such a figure can be considered as only a rough estimate. Their calculations suggest that the stored energy is proportional to the flow stress and the strain, because they observed linear hardening. Langford and Cohen could not predict the variation of $\bar{d}$ with strain because of the limitation of their theory. The energy of a cell wall cannot be independent of strain.

All theories discussed in the foregoing predict that the value of E_s/E_w is small. As will be shown below and as was pointed out by Nabarro, Basinski and Holt,[241] all theories based on dislocation interactions that predict the true rate of strain hardening also predict that E_s/E_w is small, namely of the order of 5%.

The theories are not in satisfactory accord with the observed dependence of the stored energy on strain. There are two possible reasons for this. The chief reason is that the predicted rate of increase of the dislocation density with strain is too strong a function of the strain. Strain hardening in polycrystalline metals is not generally parabolic: the flow stress is a weaker function of strain than $\bar{\sigma} \propto \bar{\varepsilon}^{\frac{1}{2}}$ and the dislocation density increases with strain more weakly than linearly. In b.c.c. metals at larger strains the stress–strain curve appears to be linear, but few stored energy values have been measured for these conditions. The second reason that the theories do not give the observed dependence of the stored energy on strain is that expressions for the strain energy of dislocations can only be approximate. This problem will be considered next.

6.3.1.2. *Stored Energy and Dislocation Structure (Substructure)*

Bailey and Hirsch[107] were the first to provide experimental evidence that the stored energy attributable to dislocations is proportional to the dislocation density. They were also the first to find that the flow stress is proportional to the square root of the dislocation density and hence that the stored energy is proportional to the square of the flow stress.

In their experiments, Bailey and Hirsch deformed silver by extension and measured the stored energy and the flow stress; they combined these measurements with measurements of the dislocation structure by transmission electron microscopy. Since the stress–strain curves of the specimens used for the stored energy measurements and of the foils used for electron microscopy were the same, Bailey and Hirsch assumed that, at a given strain, the dislocation structure was the same in both kinds of specimen. They obtained dislocation densities at each strain from transmission electron micrographs and a measure of the thickness of the foil. They believed that the values of the dislocation density were reproducible to ±25% but that the values may have been consistently low by up to ±20% because dislocations might have been lost during preparation of the foils and also because some of the dislocations might not have been in diffraction contrast. Table 6.1 presents the average dislocation densities and the stored energies obtained by Bailey and Hirsch. The electron micrographs revealed a strongly developed cell structure, with almost all of the dislocations being in the cell walls. The cell walls occupied about one-fifth of the total volume of the metal. As may be seen from Table 6.1 an appreciable fraction of the stored energy ($E_p = E_s - E_r$) was released during recovery (or prerecrystallization processes). No accompanying changes were detected in the arrangement or density of the dislocations.

TABLE 6.1

Measured Dislocation Densities and Stored Energies in Deformed Silver

True strain, $\bar{\varepsilon}$	Dislocation density, N_d, cm/cm³	Stored energy, cal/g-atom		Energy per dislocation, eV/atom plane	
		E_r	E_s	E_r/N_d	E_s/N_d
0.11	2.2×10^{10}	1.25	2.7	4.2 ± 1.0	9.1 ± 2.5
0.20	5.2×10^{10}	2.8	5.1	4.0 ± 1.0	7.3 ± 2.5
0.28	6.8×10^{10}	4.9	7.7	5.3 ± 1.0	8.4 ± 2.5

Ref.: Bailey and Hirsch.[107]

The experimental evidence of Bailey and Hirsch that the flow stress is proportional to the square root of the dislocation density has been augmented by subsequent investigations. Data in support of a relation

$$\tau = \alpha \mu b N^{\frac{1}{2}} \tag{6.3}$$

where τ is the flow stress in shear have been compiled by Otte and Hren.[270] The constant α is approximately 0.5. Further, most theories of the flow stress give

$$\tau = \frac{\mu b N^{\frac{1}{2}}}{w} \tag{6.4}$$

where w is a constant between 3 and 5.[241]

As discussed in section 5.5, the results of several investigations of the stored energy imply that the stored energy is proportional to the square of the flow stress as found by Bailey and Hirsch. The investigations in which this has been found are listed in Table 5.1. It can be concluded that for the metals and conditions of these investigations the stored energy E_s is proportional, over the ranges of strain investigated, to the dislocation density. Theories that relate the stored energy to the dislocation density will now be reviewed.

As has already been mentioned, Koehler[229] and Seitz and Read[231, 231a] found the stored energy to be proportional to the dislocation density. Seitz and Read took the energy of N dislocations to be equal to N times the energy of one dislocation. However, the value of 1 eV per atom plane assumed by them is low compared to the values obtained experimentally by Bailey and Hirsch (Table 6.1), and low also compared to calculated values.[19, 22]

Koehler by a calculation of the strain energy of dislocations arranged regularly on a

Taylor lattice[229] or as a collection of dipoles[230] found that the strain energy was given by

$$E = N \frac{\mu b^2}{4\pi} \left(\frac{m}{m-1}\right) \ln \frac{R}{2r_0} + \text{other terms} \quad (6.5)$$

where R is the distance between dislocations in the Taylor lattice or the distance between dislocations in a dipole, r_0 is the radius of the dislocation core and m is a numerical factor.

Stroh[266] made the first calculation of the strain energy of the dislocations in a pile-up. He derived the following equation for the self-energy of a single pile-up of n edge dislocations of unit length:

$$E_1 \approx \frac{n^2 \mu b^2}{4\pi(1-v)} \ln \frac{4\pi(1-v)\bar{\sigma} e^{\frac{1}{2}} R}{n\mu b} \quad (6.6)$$

where $\bar{\sigma}$ is the applied stress that piles up the dislocations and R is the size of the crystal. He assumed that removal of the stress $\bar{\sigma}$ did not alter the distribution of dislocations in the pile-up because they were somehow locked as postulated by Mott.[267] Johnson[272] has shown that Stroh's approximate result is equivalent to assuming a continuous distribution of the dislocations in the pile-up.

Stroh also calculated the energy of interaction between groups of piled-up dislocations. He argued that only the interaction between groups at either end of a slip line of length $2L$ made a contribution to the energy. By considering n dislocations in a pile-up to be equivalent to a superdislocation with Burgers vector nb he computed an interaction energy

$$E_2 \approx -\frac{n^2 \mu b^2}{4\pi(1-v)} \ln \frac{R}{2L} \quad (6.7)$$

The energy attributable to one pile-up in the dislocation structure considered by Stroh, therefore, is given by $E = E_1 + E_2$, or

$$E = \frac{n^2 \mu b^2}{4\pi(1-v)} \ln \frac{8\pi(1-v)e^{\frac{1}{2}}\bar{\sigma} L}{n\mu b} \quad (6.8)$$

The number of dislocations in a pile-up and the slip distance L are, in Mott's theory, independent of strain; thus the total energy of the entire dislocation structure is given by

$$E = N \frac{n\mu b^2}{4\pi(1-v)} \ln \frac{8\pi(1-v)e^{\frac{1}{2}}\bar{\sigma} L}{n\mu b} \quad (6.9)$$

It can be seen that the energy is proportional to the dislocation density N.

Bailey and Hirsch[107] considered the values of the stored energy of one dislocation shown in Table 6.1 in relation to the following expression for the energy of a single dislocation:

$$E = \frac{\mu b^2}{4\pi K} \left(\ln \frac{R}{r_0} + 1\right) \quad (6.10)$$

where the second term accounts for the energy residing in the dislocation core. K has the value of unity for a screw dislocation and of $(1-v)$ for an edge dislocation. Equation (6.10) gives the energy of a dislocation with a stress field extending a distance R. Bailey and Hirsch recognized that the value of R was uncertain: it could be as small as the length of a dislocation segment (200 A.U. in the specimens of silver investigated by them) or as large as the dislocation cell size (approximately 1 µm in their investigation). Trying in turn these values of R and also an intermediate value of 1000 A.U. they found that the calculated energies were smaller than the level of the values derived from experiment (column 6, Table 6.1). Bailey and Hirsch, therefore, examined whether the excess energy of a dislocation could arise from the interaction with another dislocation with Burgers vector of the same sign at a distance of 200 A.U. They took R as 1000 and 10,000 A.U. The results together with that of the calculation assuming no interaction energy and additional calculations considering the interaction energy of three or four dislocations 200 A.U. apart are given in Table 6.2. Bailey and Hirsch concluded that the stored energy can be accounted for by the self-energy of a dislocation plus an interaction energy between dislocations with Burgers vectors of like sign in groups of three or four separated by about 200 A.U. The stress field of these dislocations extends over the region of the cells ($\simeq 1$ µm). The expression used by Bailey and Hirsch to calculated the stored energy due to dislocations was

$$E = N \frac{\mu b^2}{4\pi K} \left(\ln \frac{R}{r_0} + 1 + 3 \ln \frac{R}{r} - \ln 2 - \frac{1}{2} \ln 3 \right) \tag{6.11}$$

where $r = 200$ A.U. and $R = 1$ µm.

In the same paper Bailey and Hirsch attempted to account for the stored energy as arising from groups of piled-up dislocations. They treated a pile-up of n dislocations as equivalent to a superdislocation of Burgers vector nb and on this basis derived for the energy of a pile-up

$$E = \frac{\mu (nb)^2}{4\pi K} \ln \left(\frac{R}{r_0} \right) \tag{6.12}$$

Equation (6.12) should be compared with the following modified version of eq. (6.5) representing Stroh's result for the self-energy of a single group:

$$E = \frac{\mu (nb)^2}{4\pi (1 - v)} \ln \frac{Re^{\frac{1}{2}}}{L} \tag{6.13}$$

where L is the slip distance. As shown by Seeger and Kronmüller[273] eq. (6.12) due to Bailey and Hirsch gives values too large by a factor of about 4 because the strain energy residing at small distances from the piled-up dislocations is overestimated. The conclusions regarding the occurrence of piled-up dislocations in cold-worked metals reached by Bailey and Hirsch on the basis of stored energy measurements must be disregarded.

Clarebrough et al.[274] pointed out that the equation used by Bailey and Hirsch for the energy of piled-up groups was incorrect. They calculated the energy stored in piled-up groups by Stroh's equations and the relation between dislocation density and flow stress. The values of the stored energy were found to be insensitive to the number of dislocations n assumed to be in a piled-up group, if $n > 20$. Clarebrough et al. took $n = 25$ and showed that the calculated values of the stored energy due to piled-up dislocations agreed well with measured values of the stored energy, including those of Bailey and Hirsch.[107] Thus the presence of piled-up groups of dislocations in cold-worked polycrystalline metals could not be ruled out by stored energy measurements.

Seeger and Kronmüller[273] calculated the stored energy due to edge dislocations piled up in groups. The model of the dislocation structure was that used in Seeger's theory of work hardening in Stage II of single crystals.[364] They calculated the stored energy as:

$$E_s = N \frac{\mu b^2}{4\pi(1-v)} \left(\ln \frac{R}{r_0} + 0.776n \right) \qquad (6.14)$$

The outer cut-off radius is half the distance between groups of dislocations with Burgers vectors of opposite sign. The first term is the energy of single dislocations, and the second the interaction energy between dislocations in a pile-up. Using the relation between the flow stress and the dislocation density and eq. (6.14) Seeger and Kronmüller found reasonable agreement between calculated values of the stored energy and the values reported by Bailey and Hirsch for polycrystalline silver.

TABLE 6.2

The Energy/Atom-plane of Dislocations; Alone (E_1) and Interacting in Groups of Two, Three and Four (E_2, E_3 and E_4)

R (A.U.)	E_1 (eV)	E_2 (eV)	E_3 (eV)	E_4 (eV)
200	2.66	—	—	—
1000	3.38	4.1	4.62	5.16
10,000	4.4	6.16	7.7	9.3

Ref.: Bailey and Hirsch.[107]

Moore and Kuhlmann-Wilsdorf[275] calculated the interaction energy of dislocations in a pile-up. They concluded that the numerical factor 0.776 in Seeger and Kronmüller's expression should be replaced by a factor of 1.6.

Independently of the arrangement of the dislocations in a cold-worked metal their strain energy can always be formally expressed as the sum of a self-energy and an interaction energy. Ideally the self-energy is given by

$$E = \frac{\mu b^2}{4\pi K} \ln \frac{R}{r_0} + \text{core energy} \qquad (6.15)$$

where K lies between 1 and $(1-v)$. If R is taken as a dimension of the specimen and is large, the interaction energy is a large negative quantity. Only for the theoretical "random" distribution of dislocations is the interaction energy zero.[276, 277] Various authors have introduced a modified self-energy which contains a specified amount of interaction. The reason for this modification was the recognition that dislocations do not end inside a crystal but form loops of mean radius R. The energy of interaction between segments of the same loop can be accounted for by taking the modified self-energy as given by eq. (6.15), where R is now the mean radius of the loops. This radius can be equated to the cell size.[107] An alternative argument is that the dislocation distribution about a point is correlated out to a certain radius R. The net stress at the point due to dislocations beyond this radius vanishes. The effective radius for use in eq. (6.15) is therefore R. In a third possible modification adjacent dislocations of opposite sign are grouped in pairs or dipoles. If the mean spacing between the dislocations of such a pair is R, the energy of a single dislocation is again given by eq. (6.15). R in this modification is usually $N_1^{-\frac{1}{2}}$, where N_1 is the density of dislocations in the cell walls ($N_1 \approx 5N$). This third modification to some degree specifies the dislocation distribution.

Once an expression for the self-energy has been selected, the sign of the interaction energy has meaning for the dislocation structure. For instance, if R is taken as the cell radius, a positive interaction energy implies that dislocations inside a volume of radius R are arranged so that their stress fields reinforce each other. In such a case, the dislocations may form a pile-up. A negative interaction energy implies that the dislocations are arranged so that their stress fields tend to cancel each other.

On the other hand, if R is taken as the distance between dislocations in a dipole, a positive interaction energy implies that the dislocations have not formed into dipoles, whereas a negative interaction energy implies that the arrangement tends towards multipoles.

Let us define a positive interaction between dislocations as occurring if the strain energy of a dislocation is greater than $(\mu b^2/4\pi K)\ln(R/r_0)$, where R is the cell size, and a negative interaction as occurring if the strain energy is smaller.

If we write

$$E_s = \alpha_s \mu b^2 N \qquad (6.16)$$

then

$$\alpha_s \begin{cases} > \dfrac{\ln R/r_0}{4\pi K} \\[6pt] = \dfrac{\ln R/r_0}{4\pi K} \\[6pt] < \dfrac{\ln R/r_0}{4\pi K} \end{cases}$$

depending on whether the interaction is positive, zero or negative.

From eq. (6.3), $\tau = \alpha \mu b N^{\frac{1}{2}}$ it follows that

$$\alpha_s = \frac{\alpha^2 E_s \mu}{\tau^2} \qquad (6.17)$$

Values of $\alpha^2 E_s \mu/\tau^2$ for nominally pure metals tabulated in Table 5.1 (p. 119) range from about 2 to 14. Most of the values are between 2 and 3. If we take $R = 1$ μm and $r_0 = 2$ A.U., then $(\ln R/r_0)/4\pi K$ equals 0.8: the experimental values of α_s are two or three times as large as the values of α_s for no interaction. This suggests that in cold-worked nominally pure polycrystalline metals, a positive interaction occurs between dislocations, that is, the stress field is not as relaxed as it could be.

Bailey and Hirsch have shown that the degree of positive interaction is equivalent to 3 or 4 dislocations of the same sign coexisting on a slip plane at distances of about 200 A.U. Equally well the degree of positive interaction is equivalent to the interaction in groups of 10 to 20 piled-up dislocations as given by Seeger and Kronmüller[273] (eq. 6.14) and modified by Moore and Kuhlmann-Wilsdorf.[275] Thus, as suggested by Clarebrough et al.[274] and Seeger and Kronmüller,[273] stored energy measurements do not prove or disprove the existence of piled-up dislocations. Such measurements, however, show that some positive interaction occurs. Models of the dislocation structure that do not take this positive interaction into account are not likely to be satisfactory. A positive interaction can also be inferred from the relaxation of some dislocation arrangements during recover.

Some authors based their interpretation on the ratio E_s/E_w or the ratio dE_s/dE_w. This approach is more complex than the one above relating E_s to τ^2 because these ratios depend on the extent of deformation. As pointed out by Nabarro et al.,[241] any theory that attributes the flow stress to the interaction between dislocations and gives a correct value for the rate of work hardening implies that the fraction of the energy of deformation stored is small. Nabarro et al. gave the following derivation of the ratio dE_s/dE_w:

$$dE_w = \tau d\gamma = \frac{d\gamma}{d\tau}\tau d\tau$$

$$dE_s = \alpha_s \mu b^2 dN \qquad \text{from eq. (6.16)}$$

Now

$$\tau = \alpha\mu b N^{1/2} \qquad \text{eq. (6.3)}$$

Therefore

$$\frac{dE_s}{dE_w} = \frac{2\alpha_s}{\alpha^2 \mu}\frac{d\tau}{d\gamma}$$

Taking

$$\frac{d\tau}{d\gamma} = \frac{\mu}{500} \text{ and } \alpha_s = 2.5, \alpha = 0.5$$

then

$$\frac{dE_s}{dE_w} \approx 4\%$$

This derivation brings out the parallelism between the rate of energy storage and the rate of work hardening.

6.3.1.3. *The Estimation of Dislocation Densities from Stored Energy, Density and Resistivity Measurements*

Clarebrough, Hargreaves and co-workers[3, 46, 170, 174–176, 179, 197, 278] attempted to estimate the dislocation densities in several metals after cold work. They based their estimates on measured values of the stored energy, the change in macroscopic density and the change in electrical resistivity.

Clarebrough and co-workers computed the dislocation densities by dividing the measured values of the stored energy released during recrystallization (except for aluminum for which they used the total stored energy) by the elastic energy of a single dislocation with the cut-off radius R taken as 1 cm.[19] For example energies of screw and edge dislocations in copper were taken as 3.3×10^{-4} and 5.0×10^{-4} erg/cm, respectively.[174] The arbitrariness of such a cut-off radius was recognized by Clarebrough et al. in the last paper of the series,[3] in which they also considered values of $R = 200$ A.U. and 1 μm, following Bailey and Hirsch.[107] The values of the dislocation density obtained from stored energy measurements by Clarebrough, Hargreaves and co-workers ranged from about 4×10^{10} to 3×10^{11} cm/cm^3, that is, they were of the order of 10^{11} cm/cm^3. The strains $\bar{\varepsilon}$ ranged in absolute magnitude from 0.4 to 1.7. Because the energy per dislocation was taken as constant, the estimates dislocation densities increased linearly with the stored energy. For copper, nickel and silver, these estimates agreed well with dislocation densities obtained by transmission electron microscopy.[107, 108, 158]

The estimates for aluminum[175] were larger by a factor of at least five than the values found by transmission electron microscopy by Faulkner and Ham.[215] Ham[280] showed that many dislocations were lost in thinning aluminum-rich aluminum–silver alloys that had not been aged before thinning, and suggested that at least 75% of the dislocations in cold-worked aluminum ran out during thinning. Additional evidence that dislocations may be lost in preparing thin foils of aluminum is now available.[281] The discrepancy between observed and estimated dislocation densities in aluminum can be explained in this way.

Estimates of dislocation densities by Clarebrough, Hargreaves and co-workers based on changes in macroscopic density gave values slightly higher than, but in fair agreement with, those obtained from computations based on the stored energy. The average dilation due to a dislocation was taken as 1.5 at. volume per length of $a/\sqrt{2}$, where a is the lattice parameter.[282, 283]

Estimates of dislocation densities[170, 174] based on resistivity changes due to cold work were about 50 times larger than the estimates obtained from the stored energy and the change in density. The possibility that this difference might be due to the effect of stacking faults on the resistivity[284–286] was considered. To investigate this possibility Clarebrough et al.[175] compared the dislocation densities N in aluminum, copper, nickel and alpha brass (69Cu–31Zn) computed from measured values of the stored energy with dislocation densities computed from the change in electrical resistivity, $\Delta\rho$. The ratio $\Delta\rho/N$ increased in the order copper, aluminum, nickel, alpha brass, although the values for copper and aluminum were roughly equal. The stacking fault energies cited by Clarebrough et al. decreased in the order aluminum, copper, nickel, brass. The increase of the ratio $\Delta\rho/N$ with decreasing stacking fault energy appeared to confirm that the increased electron scattering due to the separation of partial dislocations might appreciably increase the resistance of a dislocation. Clarebrough et al. pointed out, however, that to attribute the whole of the difference between the theoretical and experimental values of the resistance of a dislocation to stacking faults required that faults approximately 1 atom wide (in copper and aluminum) would increase the resistivity above the value of an undissociated dislocation by at least an order of magnitude. Unfortunately, stacking fault energies are not accurately known. Also the significance of the difference of the ratios $\Delta\rho/N$ in copper, aluminum, nickel and alpha brass is not clear since, as the same authors later noted,[176] the Fermi surfaces of these metals differed greatly.

Clarebrough et al.[176] in another investigation of this problem compared the values of $\Delta\rho/N$ for silver and gold because these metals probably had very similar Fermi surfaces, but different stacking fault energies. Since the values they obtained for the ratio $\Delta\rho/N$ differed little, they concluded that stacking faults in these metals made a negligible contribution to the resistivity of a dislocation. Basinski, Dugdale and Howie[287] reached a similar conclusion based on the measured resistivity per unit area of stacking faults in gold[288] and possible widths of the stacking fault ribbons. In alloys of low stacking fault energy, however, the resistivity due to stacking faults may be important. The dislocation densities computed by Clarebrough et al. agreed with those obtained by transmission electron microscopy in the investigations of Bailey and Hirsch[107] and Bailey.[200] The resistivity of a dislocation therefore appeared to be at least an order of magnitude larger than the various estimates made before 1963.[289–291] Values computed by Clarebrough et al.[3] from their estimated dislocation densities and measured changes in resistivity ranged from 1.3 to 3.6×10^{-13} microhm-cm per line/cm^2 in silver and copper and from 5.4 to 9.4×10^{-13} microhm-cm per line/cm^2 in nickel. Rider and Foxon[292] estimated the resistivity of a dislocation in aluminum to be of the same order, namely 2.9×10^{-13} microhm-cm per line/cm^2. Their estimate was based on resistivity and transmission electron microscopy measurements. In the same way, they estimated the resistivity of a dislocation in copper as 1.6×10^{-13} microhm-cm per line/cm^2, a value which is close to that of 1.0×10^{-13} microhm-cm per line/cm^2 found at 4.2°K by Basinski and Saimoto.[293]

Brillhart et al.[154] compared the dislocation densities in shock-loaded copper estimated from stored energy, resistivity and density measurements and measured by transmission electron microscopy. They used values of Clarebrough et al.[3] of the effect of unit length of a dislocation on these quantities. Reasonable agreement was found between the various estimates of the dislocation density.

Gangulee and Bever[180] took from the literature values of the resistivity of a dislocation in silver–magnesium alloys ranging from 3.6 to 7×10^{-13} microhm-cm per line/cm^2. They found fair agreement between the dislocation densities computed from stored energy measurements and resistivity measurements.

Earlier calculations of the resistivity of a dislocation[289-291] have been supplanted by calculations by Basinski, Dugdale and Howie[287] and Ziman.[294] The results of Basinski et al. agree reasonably well with the measured resistivities of dislocations in several metals.

The dependence of the ratio $E_s/\Delta\rho$ on the identity of the metal can be estimated from the resistivity calculations of Basinski et al.[287] and the assumption that $E_s/N \propto \mu$ where μ is the shear modulus. The calculated values of $E_s/\Delta\rho$ for the metals in which the ratio was measured by Clarebrough and co-workers and Bever and co-workers (Table 5.2) vary by a factor of three. The experimental values of the ratio also vary by a factor of approximately three. Better agreement between the experimental and calculated values of $E_s/\Delta\rho$ cannot be established at this time.

The ratios $E_s/\Delta\rho$, $\Delta\rho/(\Delta d/d)$ and $E_s/(\Delta d/d)$ may not remain constant with strain (see section 5.5). As shown by Fig. 63, the dislocation density in copper cannot bear a simple linear relation to the three quantities: stored energy, density and resistivity. In silver[107] and copper,[200] and hence probably also in gold, the dislocation density appears to be a linear function of the stored energy, at least up to moderate strains. Figure 63 shows that as the strain, and hence the dislocation density, increase the increase in dilation due to adding further dislocations is diminished. Similarly the increment of resistivity per dislocation becomes smaller as the dislocation density increases, especially above strains of $\bar{\varepsilon} \approx 0.8$.

6.3.1.4. *Stored Energy and Cell Structure*

An alternative description of the cold-worked state can be given by ignoring the individual imperfections as such and considering the assemblies of dislocations. In this connection, attention has been called[4] to the various levels on which interpretations can be developed. In particular, it may be useful to think in terms of the cell or subgrain structure and to consider the energy associated with the cell or subgrain walls.

In several investigations the stored energy has been related to the subgrain size determined from the analysis of the shapes of X-ray lines.[148, 149, 171, 188, 295, 296] Averbach et al.[148] using the analysis of Warren and Averbach[225-227] deduced a subgrain size of 200 A.U. in filings of a 75Au–25Ag alloy. The corresponding energy of the subgrain boundaries, computed from the stored energy, was about 250 erg/cm^2, a reasonable value for a low-angle boundary.

Subgrain sizes obtained by Michell and Haig[149] and Michell and Lovegrove[171] for nickel grindings and nickel deformed by compression were 400 and 1200 A.U. The measured stored energies were approximately 190 and 16 cal/gram-atom, respectively, excluding the energies associated with recovery. The corresponding energies of the subgrain boundaries can be deduced as approximately 1600 and 400 erg/cm². While the latter value may be plausible, the former is too large for a low-angle boundary. Michell and Haig[149] also obtained particle sizes from their X-ray diffraction lines by using the method of Williamson and Smallman[297] in which the strain is assumed to follow a Cauchy distribution. The resulting values of the subgrain size were much larger than those obtained by them using the analysis of Warren and Averbach.

The particle size obtained from the analysis of Warren and Averbach represents the mean length, in a specific crystallographic direction, of coherently diffracting domains. The values of this particle size generally range from a few hundred to perhaps 2000 A.U. and are therefore at least an order of magnitude smaller than the cell sizes observed under the electron microscope. Thus the "particles" determined from these X-ray analyses do not correspond to the cells of transmission electron microscopy.

The analysis by Langford and Cohen[245] of the energy stored in drawn iron wire, discussed in 6.3.1.1, is an interpretation in terms of cell wall or subgrain boundary energy. They assigned a specific energy of 500 ergs/cm² to the cell walls.

6.3.1.5. *Stored Energy and Lattice Strain*

As discussed in subsection 6.1.3, various early calculations were made of the stored energy of cold work from lattice strains as determined from X-ray line-broadening. The type of analysis of Warren and Averbach permits a separation of the effects of particle size and lattice strain on the shape of X-ray diffraction lines. The energy associated with the root-mean-square lattice strain obtained from X-ray data has been computed by several investigators.

Averbach *et al.*[148] found that the strain energy calculated from the value of the root-mean-square strain obtained in this way amounted to only about 3% of the total energy stored in filings of a 75Au–25Ag alloy. These investigators calculated the elastic energy E per unit volume from the expression

$$E = \tfrac{3}{2} Y \langle \varepsilon^2 \rangle \tag{6.18}$$

where Y was Young's modulus and $\langle \varepsilon^2 \rangle$ was the mean-square strain in an effective gauge length of a few A.U.

Michell and Haig[149] carried out a similar analysis on filings of 99.6% nickel. The measured value of the stored energy was 212 cal/gram-atom, of which about 200 cal/gram-atom could be assumed to be associated with dislocations. These authors, using the same expression as Averbach *et al.*,[148] obtained values of the elastic strain energy of 24 cal/gram-atom and 82 cal/gram-atom depending on whether they assumed a Gaussian or a Cauchy distribution of strain. On either basis, the agreement with the value of 200 cal/gram-atom was poor.

Michell and Lovegrove,[171] from X-ray diffraction data for nickel deformed by compression, calculated values of the strain energy ranging from 2.3 to 140 cal/gram-atom. The best agreement with their measured value (16 cal/gram-atom) of the stored energy was obtained if the distribution of strain was assumed to be Gaussian. These authors used a formula due to Faulkner[298] which yielded values of the strain energy about twice as large as the values given by the equation used by Averbach *et al.*[148] and Michell and Haig[149] and by the equation due to Stibitz[217] used in earlier calculations[216, 218] of the strain energy.

The values of the energy computed from root-mean-square strains obtained from the analyses of shapes of X-ray diffraction lines depend sensitively on the assumed distribution of strain.[171] Michell[295] held that the poor agreement between the stored energy measured by Averbach et al.[148] and the calculated strain energy was probably due to a lack of correspondence between the actual strain distribution in their filings and the assumed Gaussian distribution. Averbach et al.[296] suggested that the strain energy calculated from the root-mean-square strains did not include the energies of subgrain boundaries.

6.3.1.6. *Dislocation Densities from Analyses of X-ray Line Broadening*

The inconsistencies between the subgrain structure deduced from analyses of X-ray diffraction data and the cell structure observed under the electron microscope have already been noted. In addition there is the difficulty of placing a physical interpretation on the elastic strain obtained from X-ray data.

Williamson and Smallman[299] suggested that the dislocation density could be estimated from both the particle size and the root-mean-square strain obtained from X-ray data. The inequality between the two values of the dislocation density can be taken to indicate the probable dislocation arrangement.[299, 300] Williamson and Smallman took the "true" dislocation density as

$$N_d = [N_{d(L)} N_{d(\varepsilon)}]^{\frac{1}{2}} \tag{6.19}$$

where $N_{d(L)}$ and $N_{d(\varepsilon)}$ were the densities obtained from the particle size and the lattice strain.

Mikkola and Cohen[30] obtained the dislocation density in the shock-loaded specimens of Cu_3Au of Beardmore et al.[153] by applying eq. (6.19). Their values were of the order of 10^{11} lines/cm^2. Brillhart et al.[154] also used this method to determine the dislocation density in shock-loaded copper. The results were in good agreement with those obtained from stored energy, resistivity and density measurements. This agreement suggests that eq. (6.19) due to Williamson and Smallman should be applied to the data available from investigations which combined X-ray and stored energy measurements.[148, 149, 171]

6.3.1.7. *Stored Energy and Recovery*

The discussion so far in subsection 6.3.1 has been concerned with the total stored energy. Pure metals release most of this energy during recrystallization, but a release of energy before recrystallization occurs in nominally pure metals deformed at room temperature, and even in 99.999% copper[104] (subsection 4.2.2). This release during recovery is generally attributed to the rearrangement and annihilation of dislocations.[3, 46, 81, 105, 107, 170, 171, 197, 302] (The prerecrystallization peak at about 250°C in nickel[46, 170, 197] will be discussed in subsection 6.3.2.) In nominally pure metals, the changes in hardness, resistivity and density during recovery are generally small, which is compatible with the assumption that the process is one of slight rearrangements of dislocations. These rearrangements must take place among the dislocations in the cell boundaries by processes such as climb, cross-slip and cutting through other dislocations.[107]

In observations on silver and copper by electron microscopy, no significant difference could be detected between the dislocation density and arrangement before and after recovery.[107, 158, 200, 264] This does not necessarily mean, however, that the release of energy during recovery is not due to the rearrangement of dislocations. Li[23] showed that slight changes in the misorientation of two regions separated by a twist boundary composed of a crossed grid of screw dislocations could be accompanied by large changes in the stored energy. Other seemingly minor rearrangements may also cause large changes in the stored energy.

Recovery in nominally pure metals may be visualized as a process in which high-energy configurations of dislocations rearrange themselves into configurations of lower energy. The reduction in the dislocation density during this rearrangement appears to be slight; the process is primarily one by which interaction energies are reduced between dislocations. For instance, Bailey and Hirsch[107] found in silver that the slope of E_r vs. τ^2 was smaller than the slope of E_s vs. τ^2 by a factor of 1.8. As has been discussed, this means that α_s in eq. (6.16) is reduced from 2.1 to 1.2. The latter value is slightly greater than the value of α_s in the absence of interaction between dislocations. During recovery the interaction energy is reduced. The reduction in root-mean-square strain during recovery as observed in X-ray investigations[148, 149] supports this interpretation because it indicates that recovery reduces the long-range stress fields of the dislocation network.

The ratio of the energy released during recovery to the total stored energy generally decreases with increasing strain (subsection 4.1.1). This decrease in the ratio E_p/E_s suggests that dynamic recovery during continued straining replaces dislocation arrangements of high energy with arrangements of lower energy.

6.3.2. *Temperature of Deformation*

If the temperature of deformation is reduced to below room temperature the amount of stored energy is increased.[90, 128, 134–138] Part of the additional energy is released on annealing below room temperature[88–91, 135, 138, 159, 187] and is largely, if not wholly, due to annealing processes involving point defects. At least part of the remainder is added to the energy released during recrystallization. For copper it has been shown that the entire remainder is released during recrystallization.[90] The increased amount of energy associated with recrystallization can be attributed to the production at low temperatures of a higher density of dislocations and perhaps also to the retention of a greater number of high-energy arrangements of dislocations. This increase in the recrystallization energy is accompanied by a decrease in the recrystallization temperature.[90]

A similar explanation involving dislocations probably also holds for the increased amount of energy released above room temperature by gold–silver alloys deformed at 78°K.[135, 159] This is suggested by the fact that the hardness at room temperature of an 82.6Au–17.4Ag alloy deformed at 78°K was appreciably higher than that of the same alloy deformed to the same strain at room temperature.

The release of energy below room temperature after deformation at 78°K generally occurs in several overlapping stages.[3, 88–91] These stages have invariably been attributed to the annealing out of point defects. The peaks in energy release observed at about 120° and 250°C in nickel have also been explained in this way[3, 46] and will be considered in this section. Such explanations have been based on activation energies determined from the kinetics of the release of the stored energy, as in the case of nickel,[304] or from the kinetics of the changes of other properties, such as the electrical resistivity. It may also be possible to distinguish and even identify annealing mechanisms by considering the ratio of two property changes during an annealing stage, such as $\Delta E_s/\Delta\rho$ (see section 5.5).

The types and arrangements of point defects generated in face-centered cubic metals by cold work have been reviewed by Balluffi *et al.*[186] These authors considered the following annealing mechanisms for point defects as possible:

(1) migration of dispersed point defects to dislocations;

(2) clustering of like point defects;
(3) mutual annihilation of vacancy-type and interstitial-type defects;
(4) break-up of point-defect clusters and subsequent annealing of the isolated defects;
(5) trapping of defects at impurities or their release from traps.

In view of the conflicting experimental findings, Balluffi et al.[186] concluded that it was difficult to judge the significance of many of the reported activation energies. Activation energies have often been derived over a narrow temperature range within a broad and continuous annealing spectrum. It cannot be decided whether an activation energy derived in this manner refers to a simple process involving a single type of defect. The width of the annealing peaks suggests that, in most cases, the process is not a simple one. At present, the data, even for copper, the most extensively investigated metal, do not by themselves permit the unambiguous identification of a specific annealing peak with a specific mechanism. The situation does not appear to be clearer for other metals.

6.3.2.1. *Annealing of Point Defects*

In the annealing of copper after cold work or irradiation the following major stages have been identified:[89]

$$\begin{align}
&\text{I} & &< 70°\text{K} \\
&\text{II} & &70°-240°\text{K} \\
&\text{III} & &240°-330°\text{K} \\
&\text{IV} & &370°-470°\text{K} \\
&\text{V} & &> 500°\text{K}
\end{align}$$

Stage I is found after irradiation but not after cold work. Stage V is recrystallization. Stage II and Stage III may comprise several substages.[91] Although the identification of the mechanisms operating in the various stages has not yet been settled, all stages below V are generally interpreted as involving the migration of point defects.[306]

The peak observed by Henderson and Koehler[88] at about 250°K can be identified as Stage III; the group of peaks below 200°K represents Stage II. Henderson and Koehler suggested that the Stage III peak was associated with the annealing of single vacancies. They tentatively attributed the multiple peaks of Stage II to the annealing of divacancies. They suggested that the presence of three peaks in the release of the stored energy in this stage, compared with one observed in the recovery of the resistivity,[307] might arise from the effect of residual stresses on the motion of divacancies in the massive specimens used for their investigation of the stored energy.

Loretto et al.[90] identified the low-temperature annealing peaks they observed in copper as Stages II (below 200°K) and III (at about 250°K). The ratio $E_s/\Delta\rho$ was roughly constant (350 ± 60 cal/gram-atom-$\mu\Omega$cm) throughout the range of the two stages. Loretto et al. believed that the value of $E_s/\Delta\rho$ fitted a mechanism in which interstitial-vacancy annihilation was the main process operating in these stages.

In copper (Fig. 48) van den Beukel[89,91] distinguished Stages IIa, IIb, IIc, II* and III in order of increasing temperature. In binary alloys of copper with 1 at. % silver and 1 at. % gold he also recognized a Stage II** between II* and III, and a Stage III* above Stage III. The close correspondence between the peaks he obtained with nickel and copper has already been noted (subsection 4.1.2, Table 4.1). The identifications which van den Beukel assigned to the peaks in copper, silver, gold and nickel are shown in Fig. 48 and Fig. 49. It can be seen that Stage II* was not observed in silver and gold, nor IIc in gold.

Van den Beukel[91] presented a general scheme of interpretation of the various recovery stages he identified. He attributed Stages IIa and IIb to the dissociation of two different kinds of interstitial pair, IIc to the migration of interstitials to dislocations and II* to the release of interstitials trapped by impurity atoms. Stage II**, which was observed only in copper alloyed with 1 at.% silver or 1 at.% gold, was explained as the release of interstitials trapped by pairs of impurity atoms. Stage III was attributed to the release of interstitials bound to dislocations, and Stage III* to the release of interstitials bound to dislocations and impurities. Except in Stage IIc, vacancies played an important role as sinks for the released interstitials. Van den Beukel presented only slight evidence in support of these suggested mechanisms.

Van den Beukel did not observe Stage III in nickel below room temperature (Fig. 49).[89, 91] Clarebrough et al.[170] observed two recovery peaks above room temperature in 99.85% and 99.96% nickel cold worked at room temperature (Fig. 54 and Fig. 55). As mentioned in subsection 4.1.2, the lower of these peaks, centered at about 120°C, may represent Stage III recovery. The upper peak at about 250°C can be identified as Stage IV. Clarebrough et al.[46, 189] suggested that this peak corresponded to the disappearance of vacancies. Bell and Krisement[109] adopted a similar interpretation. Clarebrough et al.[3] also considered the possibility that the peaks at 120° and 250°C might be due to the annealing of single vacancies and vacancy clusters, respectively. The magnitudes of the changes in energy, resistivity and density in these stages are not inconsistent with these interpretations. Bell[111] suggested that Stage III was associated with the diffusion of interstitial atoms to immobile vacancies and impurity sites. He assumed these reactions occurred simultaneously; the first determining the kinetics and the second the intensity of Stage III. Bell also suggested that Stage IV was due to the coalescence of single vacancies into vacancy clusters.

The release spectrum of aluminum below room temperature found by van den Beukel differed markedly from the spectra of nickel and the noble metals (Fig. 48 and Fig. 49). He suggested that the prominent peak at about 210°K was Stage III and that Stage II might lie below 90°K. Lugscheider and Wildhack[92] identified the two stages of energy release found by them in aluminum cold worked at 78°K as Stages IIb and III. These stages extended from 100° to 130°K and from 130° to 250°K, respectively. They attributed Stage IIb to a recovery process involving the movement of divacancies and Stage III to the migration of interstitial atoms to vacancies.

Chuang and Bever[187] identified the annealing peaks in gold deformed at 78°K as Stages II, III and V. From the measured values of $\Delta E_s/\Delta \rho$ (section 5.5) they suggested that Stage II was associated with the annealing of interstitials or interstitial-vacancy pairs and Stage III with vacancies or divacancies.

Investigations of the energy stored during deformation at 4.2°K[138, 143a] indicated that the ratio of stored to expended energy E_s/E_w was high. The energy stored in an 82.6Au–17.4Ag alloy at 4.2°K was between three and four times that stored at 78°K. Most, but not all, of this additional energy was released by annealing at 78°K. The additional energy stored at 4.2°K can therefore be largely, if not wholly, attributed to point defects. These defects are preserved because of the reduced effectiveness of thermally activated annealing processes at this low temperature.[138] The excess of stored energy in specimens deformed at 4.2° and annealed at 78°K over that in specimens deformed at 78°K[138] may be due to a larger concentration of point defects, or it may be due to differences in the density or arrangement of dislocations. The effects of secondary deformation at 78°K following primary deformation at 4.2°K (see 3.1.5.2) suggest that differences in the density or arrangement of dislocations account for part of the energy difference.

Erdmann and Jahoda[143a] observed that the change of the thermal conductivity with strain in copper–nickel alloys was the same after deformation at 4.2°K and after deformation at room temperature. Also, the thermal conductivity of these alloys deformed at 4.2°K did not change on annealing at room temperature. Erdmann and Jahoda concluded that the dislocation substructure formed at 4.2°K was similar to that formed at room temperature, and that the large values of the stored energy at 4.2°K were chiefly due to the retention of point defects. The increased magnitude of the recrystallization peak observed with copper deformed below room temperature[90] discussed earlier in this subsection does not accord with this interpretation Also the work-softening phenomenon observed in a gold–silver alloy[159] (see 6.3.5.2) indicates that dislocations play a role in increasing the stored energy at low temperatures of deformation. It appears, however, that their contribution becomes progressively less important as the temperature is lowered.

6.3.3. *Rate of Deformation*

Any general statement about the dependence of the stored energy on the rate of deformation save a provisional one would be premature since the number of investigations concerned with this question is small. For deformation at room temperature, it is probable that the stored energy increases with increasing rate of deformation, at least to the point where adiabatic heating of the specimen is sufficient to activate recovery processes.

The increase in stored energy with increasing rate of deformation reported by Titchener and Bever[128] (Fig. 17) appeared to be connected with an increase in the density of dislocations, since the hardness also increased. The decrease in stored energy at drawing speeds above 216 in./min was not accompanied by a decrease in hardness so that the recovery processes occurring at the highest drawing speeds presumably consisted of small rearrangements of dislocations such as were discussed in 6.3.1.7.

The values of the stored energy after deformation at 4.2°K measured at 78°K by Appleton and Bever[138] indicate that, at very low temperatures, the stored energy can be a sensitive function of the strain rate. Since the changes in expended energy with strain rate were small, the changes in stored energy may be attributed to point defects. The decrease in the stored energy at strain rates above $\dot{\gamma}_{max} = 0.3$ (Fig. 18) may be explained by the annealing out of point defects under increased thermal activation.[138] The minimum in the value of the stored energy at $\dot{\gamma}_{max} = 0.2$ remains unexplained.

6.3.4. *The Deformation Process*

The stored energy data for gold–silver alloys, presented in part in Fig. 7, show that the structure of a cold-worked metal depends on the deformation process. The way in which the dislocation structures resulting from different deformation processes differ is not known at present. However, in subsection 6.2.4 it was suggested that increasing inhomogeneity of deformation may result in greater average misorientations across cell walls and hence greater energies associated with them. This may account at least in part for the trend in Fig. 7. Processes in which the deformation involves no redundant work, such as torsion and compression, appear to produce the same dislocation structures and hence equal stored energies (Fig. 10).

Mikkola and Cohen[301] examined by X-ray diffraction and optical and transmission electron microscopy specimens of Cu_3Au shock-loaded by Beardmore *et al.*[153] They found that the chief difference between shock loading and conventional deformation of ordered Cu_3Au was that shock loading disordered the alloy more extensively. The antiphase boundary probability as a function of strain found by Mikkola and Cohen for shock-loaded and conventionally deformed ordered Cu_3Au is shown in Fig. 65. The marked

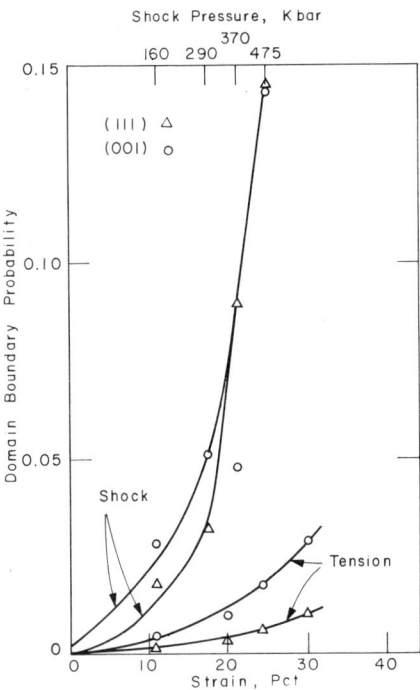

FIG. 65. The (111) antiphase domain boundary probability in Cu_3Au and the (001) antiphase domain boundary probability in Cu_3Au versus transient strain in shock loading and strain in tension. Mikkola and Cohen.[301]

increase in the antiphase boundary probability between transient strains of 15 and 25% (corresponding to pressures ranging from 290 to 370 kbar) can explain the marked increase in stored energy over the same strain range shown in Fig. 20. The long-range-order parameter was found to decrease rapidly at shock pressures somewhat above 370 kbar. Mikkola and Cohen observed deformation twinning in shock-loaded specimens but not in tensile specimens. They also found that the dislocation density determined by X-ray diffraction in shock-loaded specimens was only 30% greater than that in specimens deformed in tension to a permanent strain equal to the transient strain in shock loading. The energy due to dislocations and twins was calculated to be only a small fraction of the total stored energy. Mikkola and Cohen suggested that antiphase boundaries did not contribute to the flow stress of Cu_3Au, which would explain why the ratio of stored energy to hardness was greater in shock-loaded Cu_3Au than in conventionally deformed Cu_3Au.

Brillhart et al.[154] found that the dislocation density measured by electron microscopy in oxygen-free high-conductivity copper shock loaded to 345 and 435 kbar was approximately 10^{10} cm/cm^3, whereas the dislocation density calculated from stored energy values would be approximately 10^{11} cm/cm^3, if the stored energy were attributed to dislocations and coherent twin interfaces. The value of 10^{10} cm/cm^3 for the dislocation density is of the order observed by Bailey,[200] using electron microscopy on copper specimens at a permanent strain equal to the transient strain corresponding to a shock pressure of 345 to 435 kbar. Brillhart et al. concluded that the increased stored energy of shock-loaded copper was due to twin interface energy, that the twin interfaces were incoherent, and hence that the twin interface energy was of the order of a few hundred ergs/cm^2.

The inconsistency between the annealing kinetics of copper as observed by Brillhart et al. and those observed by Iyer and Gordon[151] has not yet been explained.

6.3.5. Deformation History
6.3.5.1. Bauschinger Effect

The decrease in hardness found by Iyer and Gordon[157] during secondary compression after primary extension of copper (see 3.1.5.1) indicated that the associated decrease in the stored energy was due to the rearrangement and annihilation of dislocations. This decrease is a manifestation of the Bauschinger effect. Since the secondary compression affected the amount of energy associated with recovery, as well as that associated with recrystallization, the recovery process presumably involved the rearrangement and annihilation of dislocations.

Hargreaves et al.[158] considered several possible interpretations of their observations of the changes in stored energy during secondary compression after primary extension (3.1.5.1, subsection 4.1.5 and section 5.4). They concluded that the changes in energy and density during recovery could not be explained by annihilation of dislocations or annealing out of point defects, in particular vacancies, since about half of the energy associated with recovery was released without a measurable change of density. Also, the ratio $\Delta E_s/\Delta d/d$ for the recovery stage as a whole was too large to be explained by the annealing of vacancies. Hargreaves et al. suggested that the most likely mechanism for recovery was the relaxation of high-energy configurations of dislocations. They pointed out that their explanation was compatible with the suggestion of Bailey and Hirsch[107] that the recovery energy resided in small groups of two to four dislocations. Like Bailey and Hirsch, however, they could not detect by electron microscopy a change in the number and arrangement of dislocations during recovery.

Hirsch[308] argued that the changes in the stored energy observed by Hargreaves et al. during recovery could not be due to changes in the elastic strain energy of dislocations if no change occurred in the macroscopic density. He proposed that the change in energy during the early stages of recovery could be explained by a mechanism in which the vacancies migrated to form progressively larger clusters, which would cause little change in density. He explained the later stage of recovery, in which the density increased appreciably, as the collapse of vacancy aggregates into dislocation loops followed by coarsening and annealing of these loops.

While the mechanism proposed by Hirsch may be the correct one, the absence of a change in density does not appear to preclude changes in the number and arrangement of dislocations in a complex dislocation substructure. According to Zener[309] the average dilatational strain $\bar{\theta}$ of a crystal lattice may be expressed as

$$\bar{\theta} = k_d W_d + k_s W_s \tag{6.20}$$

where W_d and W_s are the dilatational and shear work of elastically deforming the lattice and k_d and k_s are constants. The elastic strain energy of the dislocation is the sum, $W_d + W_s$. The quantities W_d and W_s depend on the kind and arrangement of dislocations present. A unique relation between the density and the elastic strain therefore cannot exist.

6.3.5.2. Work Softening

The work softening observed by Titchener and Bever[159] (see 3.1.5.2) occurred during a secondary deformation of the same kind as the primary deformation, but carried out at a higher temperature. Since both the hardness and the stored energy decreased, the mechanism may be presumed to involve the rearrangement and annihilation of dislocations. The striking changes in the dislocation substructure observed by Keh and Weissmann[240] (see 6.2.5.2) in iron subjected to a similar two-stage deformation are of interest in this connection.

The results of Erdmann and Jahoda[145] on a copper–nickel alloy point to an energy release as in work softening at temperatures as low as 4.2° to 70°K. On the other hand an energy release as in work softening was not seen by Appleton and Bever[138] when the second deformation was at 78°K.

The occurrence of work softening is probably related to the effect of a higher temperature producing an increase in dislocation mobility off the slip plane and hence allowing a lower energy configuration to be achieved and annihilation to occur at the higher temperature under flow stresses of the same order as at the lower temperature. The processes could be cross slip or conservative climb.

6.3.6. Cyclic Deformation (Fatigue)

Most of the energy stored in 99.98 % copper during cyclic deformation is released by a recovery process[161] (subsection 4.1.6). This contrasts with the behavior in deformation by compression, tension or torsion of copper of this purity which releases most of the stored energy during recrystallization.[46] Clarebrough et al.[161] suggested that much of the energy stored by cyclic deformation resided in loosely bound dislocations which were eliminated at relatively low temperatures.

Segall, Partridge and Hirsch[310] showed by electron microscopy that cyclic deformation at small strain amplitudes produced large numbers of elongated dislocation loops, and that in copper these loops annealed out in the same temperature range as the release of the stored energy. They estimated that the amount of energy released by the disappearance of all the observed loops was equal to 0.95 cal/gram-atom. The energy released by specimens cycled in reversed tension at a stress amplitude of 10,000 psi was 1.9 cal/gram-atom.[161] The energy released during recovery by specimens loaded in reversed bending at a stress of 25,000 psi was appreciably greater than this; Clarebrough et al.[3] pointed out that annealing of dislocation loops therefore could not account for all the stored energy released during recovery of specimens subjected to cyclic loading at high stresses. The absence of recrystallization in specimens cycled at small stress amplitudes is understandable; most of the dislocations are in loops that anneal out in recovery.

Broom and Molineux[311] suggested that cyclic straining might produce large concentrations of vacancies. On this basis, Broom and Ham[312] put forward an explanation of cyclic strain softening[161] (subsection 3.1.6). They suggested that vacancies might migrate to dislocations in such numbers as to produce climb, which would permit rearrangement of dislocations and consequent softening. With nickel deformed in fatigue, however, Clarebrough et al.[161] observed no large release of energy near 250°C, the temperature at which a peak occurred after deformation by compression or torsion (Fig. 45). Since this peak has been explained by the annealing out of vacancies (see 6.3.2.1), cyclic straining appears to preserve no abnormal concentration of vacancies.

Feltner and Laird[242] offered an interpretation of the change in stored energy over a half cycle as observed by Halford.[69] They argued that the cyclic stress amplitude was large and postulated that plastic straining occurred by the bowing out of dislocations between obstacles according to the model suggested by Kuhlmann-Wilsdorf.[313] They could account for the change in the stored energy over a half cycle by the superposition of two events: (1) repulsive interactions between a cell wall and a dislocation forced against it, which caused the stored energy to build up towards the end of a cycle, and (2) annihilation of about half of the total length of the mobile dislocations at the very end of a cycle when the dislocations came into contact with the cell walls. Both of these events contribute to the rapid energy release at the end of a half cycle observed by Halford.

6.3.7. Grain structure

In polycrystalline metals the grain boundaries form the main obstacles to slip at small strains. The closer spacing of boundaries in fine-grained compared with coarse-grained copper can account for the values of the flow stress, hardness, and stored energy being higher in the former as observed by Clarebrough et al.[124] at small strains. Deformation near grain boundaries is more complex than in the interior of grains and this complex deformation occurs in a larger volume fraction of fine-grained than coarse-grained specimens. Clarebrough et al. suggested that, at the strain at which the hardness and stored energy of the coarse-grained copper attained the same values as in the fine-grained copper, the cell sizes and concentrations of dislocations in the cell walls were essentially equal.

The recrystallization temperature of the fine-grained specimens was lower than that of the coarse-grained specimens at small strains. Clarebrough et al. explained this as due to a greater density of dislocations in the former. If the density and arrangement of dislocations within grains were similar in the coarse- and fine-grained material at large strains, the difference in recrystallization temperature persisting at these strains would have to be due to other factors. Clarebrough et al. suggested that the larger total area of grain boundaries in the fine-grained material contributed to the lowering of the recrystallization temperature. They also believed that the heat treatments used may have caused the impurities to be more highly segregated at the boundaries of the coarse-grained than the fine-grained specimens and therefore to be more effective in immobilizing at the boundaries the dislocations required for the growth of a nucleus. They also attributed the absence of a recovery peak and the greater breadth of the recrystallization peak in the coarse-grained material to more pronounced segregation of impurities.

According to an explanation of the dependence of the stored energy on grain size advanced by Conrad and Christ[314] the stored energy is related to the strain $\bar{\varepsilon}$ and the grain size d by

$$E_s = (\bar{\varepsilon}/kb)(1/d^n) \tag{6.21}$$

where k and n are empirical constants that depend on the strain. Conrad and Christ attempted to demonstrate that n, determined from the results of Clarebrough et al.[124] for copper, followed the same strain dependence as that determined from the dislocation density in iron;[240] in this they assumed a linear relation between the stored energy and the dislocation density. They took the values of k and n determined from the data on iron to calculate the dislocation densities in copper[46] and silver.[107] Their calculated values agreed well with dislocation densities determined from observations by electron microscopy.[200, 107]

There seems to be no doubt that the larger values of the stored energy in fine-grained as compared with coarse-grained metals at small strains are due to a higher dislocation density at the same strain (see subsection 6.2.7). Additional support for this conclusion is provided by the findings of Williams[54] and Wolfenden[59] that curves of stored energy versus the square of the flow stress are independent of the grain size.

6.3.7.1. Single Crystals

All investigators except Krivobok[93] and Taoka et al.[97] used single-step methods to measure the energy stored in cold-worked single crystals (see Table 3.1). Krivobok did not report values of the stored energy. Taoka et al. used rolling to large strains as the deformation process. Consequently all published values of the energy stored in single crystals deformed into Stage I or Stage II of the stress–strain curve were determined by single-step methods.

Values of the stored energy measured by single-step methods are generally higher than values measured by two-step methods. This was pointed out in section 2.1 and subsection 3.1.1, and the possibility that it results from systematic errors in single-step methods was discussed. Also the stored energy measured by a single-step method is larger than the energy measured by a two-step method by the amount of energy released immediately after the deformation. The difference may amount to 5–25% of the total energy.[38,43,51–54] In interpreting the energy stored in single crystals of copper and silver, Wolfenden has suggested that point defects contribute approximately $2.3\gamma^2$ cal/gram-atom, where γ is the shear strain.[55, 58] In deriving this expression, Wolfenden used an expression proposed by Friedel[20] for the number of point defects generated during plastic deformation. Over the range of strains to which single crystals have been deformed, the calculated contribution to the energy by point defects is approximately 20 to 50% of the total stored energy.[55–58, 315] Kuhlmann-Wilsdorf[316] and Kronmüller and Wilkens[317] accepted the view that there is a contribution to the energy from point defects. Apart from the energy released immediately after deformation (which Williams attributed to dislocations, however) there is little evidence for the annealing out above room temperature of point defects in deformed copper and silver. The stored energy in these metals deformed at room temperature, therefore, is not likely to involve a contribution by point defects.

The stored energy in polycrystalline copper, deformed with an expended energy of approximately 125 cal/gram-atom, as determined by Williams[51] with a single-step method, was approximately twice that determined by White[125] and 60% larger than the value determined by Gordon,[104] both of whom used two-step methods. On this basis it is probable that the energy of the dislocation structure in the single crystals is not less than one half of the published values of the stored energy, and may be not less than two-thirds. From an average experimental value of dE_s/dE_w of 0.16 for single crystals of copper and silver in Stage II, we estimate that the rate dE_s/dE_w attributable to dislocations is between 0.08 and 0.16.

Seeger and Kronmüller[273] calculated the rate of energy storage dE_s/dE_w from the model of the dislocation arrangement assumed in Seeger's theory of work hardening.[364] This arrangement is one of groups of piled-up dislocations. They found that in Stage II

$$\frac{dE_s}{dE_w} = \frac{4\pi\theta_{II}(2-\nu)}{\mu}\left(\frac{1}{n}\ln\frac{R}{r_0} + 0.776\right) \qquad (6.22)$$

where θ_{II} is the work-hardening rate ($\approx 3 \times 10^{-3}$ for copper), n is the number of dislocations in a pile-up (≈ 25) and R is one-half of the distance between pile-ups. Substituting numbers in eq. (6.22), they found that dE_s/dE_w in copper and silver was approximately 0.07.

Kronmüller[262] later calculated the energy in excess of that given by eq. (6.22) that would come from dislocations produced during slip on secondary systems. He found that dE_s/dE_w should be reduced by no more than 25%. Kronmüller and Wilkens[317] used eq. (6.22) with revised numerical values to calculate dE_s/dE_w and included the energy due to secondary dislocations. They concluded that dE_s/dE_w lay between 0.068 and 0.09. This is nearly in the range of 0.08–0.16 estimated from experimental values. The main features of the interpretation by Seeger and co-workers are that R is taken as the distance between pile-ups and that there is, in addition to a self-energy of dislocations, a positive interaction energy.

As part of the "mesh-length" theory, Kuhlmann–Wilsdorf estimated dE_s/dE_w in Stage II of the deformation of single crystals as approximately 0.16.[313] In a later paper[268] she calculated dE_s/dE_w as

$$\frac{dE_s}{dE_w} = \frac{2\theta_{II}}{\pi\mu}\ln\frac{\mu}{2\tau} \qquad (6.23)$$

In the later paper the stored energy is the self-energy of single dislocations, the stress field of which extends a distance $N^{-\frac{1}{2}}$. Substituting numbers in eq. (6.23) gave dE_s/dE_w as 0.01 or 0.02.

Wolfenden[58] noted that Kuhlmann-Wilsdorf's theory resulted in values of dE_s/dE_w smaller than values measured by him with single crystals of silver. This was the beginning of a controversy involving Wolfenden, Kuhlmann-Wilsdorf and Kronmüller and Wilkens.[56,58,275,315–320] The points at issue were whether or not, in Stage II of single crystal deformation, there is a positive interaction energy between dislocations as would be involved if pile-ups formed and whether energy measurements can provide evidence regarding the interaction energy.

The calculated and experimental values of dE_s/dE_w suggest that the energy of individual dislocations with stress fields extending a distance $N^{-\frac{1}{2}}$ is insufficient to account for the total energy measured. A positive interaction energy between dislocations which is implicit in the theory of Seeger and Kronmüller seems to be required.

We can further examine the possibility of a positive interaction energy in cold-worked single crystals by considering the dependence of E_s on τ^2 observed experimentally (Table 5.1) as was done for polycrystalline metals in 6.3.1.2. Values of $\alpha_s = \alpha^2 E_s\mu/\tau^2$ (eq. 6.17) where α for single crystals is taken as 0.4[262] are tabulated in Table 5.1. The values of α_s lie between 1.7 and 8 with an average of about 5. If we take $R = 10$ μm and $r_0 = 2$ A.U., then $(\ln R/r_0/4\pi K$ (see 6.3.1.2) is equal to approximately 1, that is, approximately one-fifth of the average experimental values of α_s. Considering that for single crystals the experimental values of the stored energy obtained by single-step methods may exceed the total energy due to dislocations by a factor of about two, we conclude that the experimental values of α_s based on the energy due to dislocations only lie between $2\frac{1}{2}$ and 5, that is they are two to five times the value for zero interaction. This suggests that a positive interaction occurs between dislocations in single crystals strained into Stage II. This interaction may be produced by pile-ups as proposed by Seeger and Kronmüller. Hirsch's[321] theory of work hardening also implies a dislocation structure similar to that envisaged by Seeger.

The increasingly steep slope of the curve of the stored energy in single crystals versus strain can be readily interpreted. Since E_s is proportional to τ^2 and in Stage II τ is proportional to γ, it follows that E_s is proportional to γ^2.

Williams[50] has suggested that the curves of dE_s/dE_w and $\delta E_s/\delta E_w$ against strain show maxima. He based this suggestion on the observed relation that E_s is proportional to τ^2; consequently dE_s/dE_w is proportional to $d\tau/d\gamma$ (see 6.3.1.2). The rate dE_s/dE_w is therefore expected to be lower in Stage I than in Stage II by a factor of about 10. Williams[50] found that his results for copper single crystals followed such a trend. It is unlikely, however, that in easy glide, E_s is proportional to τ^2. Such a proportionality depends on the relation $\tau \propto N^{\frac{1}{2}}$. In easy glide the latter relation has been shown not to hold.[260, 261] Straining in easy glide fills a single crystal with dislocation dipoles and multipoles. Argon and Brydges[261] have measured the average distance between dislocations in a dipole and the rate of dislocation accumulation $dN/d\gamma$. A rough calculation using the results of these investigations shows that dE_s/dE_w falls from about 0.5 to about 0.05 in the course of easy glide strain. It would be desirable to verify the magnitude of these values experimentally.

Nakada[68] suggested that the comparatively low rate of energy storage in an aluminum single crystal compressed in the [100] direction was a result of appreciable cross slip. The rate of work hardening of aluminum crystals of this orientation in Stage III is also very low.[322]

The results of three investigations by electron microscopy of the dislocation structure in cold-rolled single crystals of silicon–iron are pertinent to the investigation of the stored energy by Taoka et al.[97] Hsun Hu[323] examined the structure of cold-rolled Fe–3Si single crystals which initially had a (110) [001] orientation. Hsun Hu[324] and Walters and Koch[325] examined the structure of cold-rolled Fe–3Si of initial (100) [001] orientation. The crystallographic texture after rolling in the three investigations consisted of two major components. Extensive deformation banding was observed in the (100) [001] oriented crystal. By a rather simple interpretation of the results of Taoka et al., the stored energy released during the recrystallization is greater in the less symmetric orientation in which more slip systems need to operate, resulting in a higher dislocation density. Because of the changes in the crystallographic texture developed by rolling, the relative stored energy values cannot be used to interpret the recrystallization textures of polycrystalline metals. However, Dillamore et al.[167] calculated the energy stored in grains of several orientations in cold-rolled low-carbon steel from the number of cell walls and the misorientation across them. Using these calculations, they analyzed the formation of recrystallization textures.

6.3.8. Purity

Impurities tend to cause an increase in the amount of stored energy. In addition they lead to striking changes in the kinetics of the release, especially during recovery. They invariably raise the recrystallization temperature.

These effects are interpreted as arising from the pinning of dislocations by impurity atoms. In the absence of impurity atoms, the rearrangement and mutual annihilation of dislocations take place more freely. Much of the recovery occurring in impure metals is therefore not observed in purer metals because it has already occurred during and immediately after deformation. The release of appreciable amounts of energy immediately after deformation has been well established by Williams.[38,43,51,54]

The recovery stage immediately preceding recrystallization in impure metals (for example see Fig. 45 and Fig. 52) represents, at least in part, the rearrangement of dislocations previously immobilized in configurations of high energy by impurity atoms during deformation. The possible role of vacancies in this stage of recovery, as suggested by Hirsch,[308] should not be overlooked.

The complexity of the effects of impurities on the stored energy was reviewed in subsection 3.2.2. As pointed out there, it is not enough to consider only the total purity of the metal. The identity and distribution of the impurity atoms appear to play an important part in affecting both the amount of energy stored in nearly pure metals and the kinetics of its release, that is, in affecting the mechanisms of energy storage and energy release. The effect of the identity of the impurities was demonstrated by investigations of Clarebrough, Hargreaves and co-workers,[46,158] White and Koyama[74] and Wenzl.[95]

The effect of impurity atoms probably is greater the more they differ from the solvent atoms. For example, arsenic caused a more marked increase in the energy stored in copper[46] than did silver.[51] Clarebrough et al.[302,90] attributed the extensive plateau in the release of energy by arsenical copper during recovery to the release of dislocations pinned by arsenic atoms. In this connection, it is significant that arsenical copper strain-ages rapidly at room temperature.[326,327] The release of energy during the recovery observed by Clarebrough et al.[46] probably was not due to precipitation since the concentration of arsenic (0.35%) was well below its solubility in copper.[328] Impurity pinning apparently did not occur in Wenzl's investigation of arsenical copper since he did not observe a recovery plateau.

The absence of a plateau of energy release preceding recrystallization in 99.85% nickel (Fig. 54) and its presence in 99.6% nickel (Fig. 45) are in accord with the disappearance of the recovery plateau with increasing purity observed in copper.[46] The recovery plateau in nickel has been attributed to the rearrangement of dislocations.[46] The recovery mechanisms are complex, however, since in nickel of higher purity, the plateau was resolved into two distinct peaks.[170]

The recovery processes occurring below room temperature after deformation at low temperature generally appear to be less sharply resolved when impurities are present.[91] According to van den Beukel[91] the effect of impurity atoms is to shift each recovery process to a higher temperature. The decreased resolution of the release spectrum presumably implies that the range of values of the activation energy associated with a given annealing process tends to be wider.

6.3.9. The Identity of the Metal

The difficulties of comparing different metals with respect to the energy of cold work stored by them have been discussed in subsection 3.2.3. As Titchener and Bever[1] pointed out, the amount of the stored energy is related in some degree to the melting point and recrystallization temperature of a metal. The investigations of Clarebrough and co-workers[124,170,174–176] and Williams[38,52] have firmly established the existence of an approximate parallelism between the amount of the stored energy and the melting point.

Holt[329] pointed out that in different cold-worked metals having similar dislocation structures the stored energy due to the dislocations should differ only by a factor μb^2, where μ is the shear modulus and b the Burgers vector of a dislocation in the lattice. He defined a normalized recrystallization energy $E_r/\mu b^2$; E_r is the stored energy released during recrystallization, which can be attributed to dislocations. He compared the quantity $E_r/\mu b^2$, computed from values of the energy measured in silver, copper, gold, nickel, and aluminum, deformed in compression in one laboratory to comparable strains. Table 6.3 presents these normalized recrystallization energies in order of decreasing magnitude. Since the annihilation of dislocations during cold work is aided by cross slip, and cross slip occurs more readily the higher the stacking fault energy, Holt suggested that the normalized recrystallization energy might be inversely related to the stacking fault energy in face-centered cubic metals. In support of his suggestion, he tabulated published values of the stacking fault energy[286, 330-332] which are included in Table 6.3.

Since Holt's publication various new values of the stacking fault energy have been reported.[333-336] With the possible exception of a reversal in the order of nickel and aluminum, however, the order of the stacking fault energies of the metals Ag, Cu, Au, Ni and Al appears to be the same as that in Table 6.3. The magnitudes of these energies are also still acceptable. It can be concluded that the unusually large stored energy of silver when it is ranked with other metals on the basis of relative melting temperatures probably results from a comparatively low stacking fault energy.

TABLE 6.3

Normalized Recrystallization Energies and Stacking Fault Energies of Several Face-centered Cubic Metals

Metal	Purity, %	Normalized recrystallization energy, $E_r/\mu b^2$	Stacking fault energy		
			γ, erg/cm^2	Reference	Source
Ag	99.98	7.65	25	330	Observations of dislocation nodes
Cu	99.98	4.48	70	330	Observations of dislocation nodes
Au	99.9	3.7	75	332	
Ni	99.85	2.54	150	331	Observations of dislocation nodes
Al	99.91	1.57	200	286	Generally accepted value

Ref.: Holt.[329]

6.3.10. Composition and Configuration of Alloys
6.3.10.1. Solid Solutions

The stored energy of cold work is considerably larger in solid solutions than in the component metals. This is partly due to the destruction of short-range order by deformation and partly to the suppression, by alloying, of dynamic recovery and of recovery immediately after deformation. The suppression of these recovery processes can be in part explained by a change in the deformation mode from diffuse slip in metals to localized slip in alloys.[337] The destruction of short-range order and the suppression of recovery after deformation can also account for the observed larger values of the ratio E_s/E_w.

The contribution of the destruction of short-range order to the stored energy of alloys was investigated by Averbach et al.[148] Applying the quasichemical theory of solid solutions, they expressed the stored energy due to the destruction of short-range order as

$$E_s(\text{SRO}) = N_0 Z v x(1 - x)\delta\alpha \qquad (6.24)$$

where N_0 is Avogadro's number, Z is the coordination number of the first shell, v is the interaction energy, x is the atomic fraction of the solute atom, and $\delta\alpha$ is the change in the short-range-order parameter with deformation;[338] the interaction energy is the bond energy between unlike atoms less one-half of the sum of the bond energies between like atoms of both components. With $N_0 v$ equal to -250 cal/gram-atom and with data on the change of the nearest-neighbour short-range-order parameter obtained by X-ray diffraction. Averbach et al. estimated that approximately 21 cal/gram-atom or 28% of the stored energy in filings of a 75Au–25Ag alloy was attributable to the destruction of short-range order.

Greenfield and Bever[136] evaluated in a similar manner the stored energy due to the destruction of short-range order in drillings of gold–silver alloys. They assumed that drilling reduced the short-range-order parameter by 50% as had been determined for filings by Averbach et al. Greenfield and Bever found that $E_s(\text{SRO})$ was largest for the alloy of equiatomic composition, for which the fraction $E_s(\text{SRO})/E_s$ was approximately 0.3.

Gangulee and Bever[180] used the quasichemical theory to calculate the stored energy due to the disordering of silver–magnesium alloys by wire drawing. They defined v and $\delta\alpha$ in eq. (6.24) as the effective interaction energy and the change in effective short-range-order parameter, respectively. Gangulee and Bever took $N_0 v$ as -930 cal/gram-atom and assumed that $\delta\alpha$ as a function of strain was given by

$$\delta\alpha = \alpha_0[1 - \exp(-B\bar{\varepsilon})] \qquad (6.25)$$

where α_0 is the initial effective short-range-order parameter and B is approximately 0.4. At a strain of unity, $E_s(\text{SRO})$ was calculated as 41 cal/gram-atom for an Ag–3.0Mg alloy, as 100 cal/gram-atom for an Ag–5.4Mg alloy, and as 119 cal/gram-atom for an Ag–7.0Mg alloy. These values are higher than those calculated for gold–silver alloys by Greenfield and Bever.[136] They range from 0.55 to 0.75 of the total measured stored energy.

The stored energy due to the destruction of short-range order may also be estimated from data obtained by annealing methods, since reordering during annealing may produce a characteristic peak of energy release. An exact separation of the stored energy due to short-range order from the total, however, is not always possible, for example, if reordering occurs simultaneously with other restoration processes.

The investigations of Clarebrough et al.[179, 191–193] dealt with the contribution of changes in short-range order to the stored energy of cold work in alpha brass. The main peak (Stage 2 in Fig. 59) of the energy release represents predominantly the return of short-range order destroyed during cold working. According to Clarebrough et al.,[193] however, it is unlikely that a single mechanism is responsible for Stage 2, since the changes in the properties investigated, namely the stored energy, hardness, electrical resistivity and density in this stage are complex. Some rearrangement of dislocations must take place.

Clarebrough et al. interpreted the two peaks of Stage 1 (Fig. 59) as vacancy-assisted reordering, with an additional contribution due to annealing out of vacancies. The ratio $E_s/(\Delta d/d)$ in Stage 1 is too large to be accounted for by a reduction of vacancy concentration only. The two peaks were explained by a distribution of vacancies between two types of site.[193] The decrease of the resistivity and the increase of the density produced by small extensions (sections 5.3 and 5.4; Table 4.3) arise presumably because small amounts of deformation increase the local order.

Clarebrough et al. showed that a considerable fraction of the stored energy in alpha brass was due to the destruction of short-range order. At the largest strain ($nd/l = 1.87$) 80% of the total stored energy was accounted for by this mechanism.

Taoka et al.[97] attributed the energy released in the first stage of annealing of Fe–2.98Si to reordering. The energy asociated with this reordering was approximately 30% of the total stored energy.

The energy attributable to the destruction of short-range order in the investigations discussed range from 25 to 80% of the total stored energy. Differences in the extent of deformation and in the deformation process in part account for this variation. Furthermore, the identity of the alloy system has a major effect; whereas the effect of the destruction of short-range order on the stored energy was relatively small in gold–silver alloys, it was relatively large in silver–magnesium alloys and alpha brass. Correspondingly, the integral heats of formation are considerably larger in absolute magnitude in alpha brass and silver–magnesium than in gold–silver at equal solute concentrations.[339]

Greenfield and Bever[136] observed that the stored energy not attributable to the destruction of short-range order increased with increasing solute concentration. Gangulee and Bever[180] obtained a similar result. The stored energy released during recrystallization in the investigation of alpha brass by Clarebrough et al. was larger than the energy stored in copper deformed by them to the same strain (Table 4.3 and Fig. 8). Taoka et al.[97] found that the energy released in the recrystallization of Fe–2.98Si was greater than that released by nominally pure iron. Taken together, these results lead to the conclusion that the dislocation density in solid solutions is greater than in metals deformed to the same strain (subsection 6.2.10). The dislocations are probably also arranged in configurations of higher energy.

During the annealing of the alpha brass investigated by Clarebrough et al.[179, 191–193] the return of short-range order, recovery and recrystallization were fairly clearly separated, at least at moderate strains. The annealing curves of Sato[73] for a 55Cu–26Ag–19Ni alloy indicate that such a separation may not occur, and that the processes of reordering, recovery and recrystallization may overlap.

6.3.10.2. *Alloys with Long-range Order*

The stored energy of cold work in alloys with initial long-range order is especially high. This is due to the creation of antiphase domain boundaries and the reduction of the long-range order by deformation. Cohen and Bever[130] found that rolling to a strain of $\bar{\varepsilon} = -0.89$ reduced the long-range-order parameter S in ordered Cu_3Au from 0.85 to 0.55. The stored energy at this strain was approximately 365 cal/gram-atom. Roessler and Bever[137] showed that the curve of the stored energy versus the change in resistivity observed for the initially ordered alloy Cu_3Au (Fig. 64) deformed at 78°K was the same as that after deformation at room temperature; this suggests that the predominant mechanism of deformation and energy storage is the same at both temperatures. Mikkola and Cohen[301] found that the antiphase domain boundary probability in specimens of Cu_3Au shock loaded by Beardmore et al.[153] increased at a high rate in the range of shock pressures over which the stored energy increased rapidly (Fig. 65). In this range of pressure, the microhardness measured by Beardmore et al. and the dislocation density measured by Mikkola and Cohen changed little. Consequently, the effects of the creation of antiphase domain boundaries and of the reduction of long-range order on the stored energy of ordered alloys are established.

Robinson and Bever[183] compared the ratio of the stored energy to the change in resistivity of TlBi$_2$ deformed to a given strain with this ratio for other intermetallic compounds deformed to the same strain. They argued that the stored energy would be a function of the number and energy of broken interatomic bonds, but the resistivity would be a function only of the number. When they plotted $E_s/\Delta\rho$ at the same strain versus the integral heat of formation of the compounds, they found that the ratio increased with the heat of formation. The curve plotted by Robinson and Bever with an additional point for the initially long-range ordered alloy Ag$_3$Mg[180] is shown in Fig. 66. Robinson and Bever pointed out that it would be desirable to obtain the corresponding data for a series of compounds of the same structure deformed at temperatures which are the same fraction of their melting points.

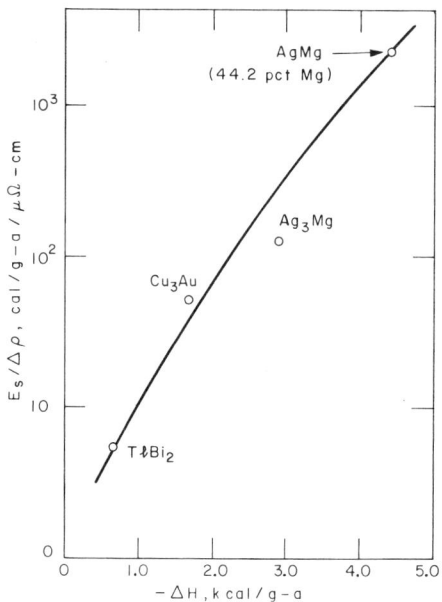

FIG. 66. The ratio of the stored energy to the change in resistivity for alloys with long-range order deformed at room temperature plotted against their heats of formation. The alloys TlBi$_2$ and AgMg were deformed in tension to a shear strain of 0.31. The alloys Cu$_3$Au and Ag$_3$Mg were deformed by wire drawing to equivalent strains. Data of Cohen and Bever,[130] Gangulee and Bever[180] and Robinson and Bever.[140, 183]

Various authors have calculated the magnitude of the energy stored by the creation of antiphase domain boundaries and the destruction of long-range order. Cohen and Bever[130] measured by X-ray diffraction antiphase boundary probabilities in deformed initially ordered Cu$_3$Au. From published values of the antiphase boundary energies they calculated the energy at a strain of $\bar{\varepsilon} = 0.26$ due to the production of antiphase domain boundaries as amounting approximately to 5% of the total stored energy. As already mentioned, Mikkola and Cohen[301] measured antiphase boundary probabilities in shock-loaded initially ordered Cu$_3$Au specimens. Using a value of 94.5S^2 erg/cm$^{2(340)}$ for the energy of a (111) antiphase domain boundary, Mikkola and Cohen found that the energy E_s(APDB) due to antiphase boundaries on (111) planes could be expressed as

$$E_s(\text{APDB})^{(111)} = 4953 S^2 \gamma \text{ cal/gram-atom} \quad (6.26)$$

where S is the long-range order parameter and γ is the probability of (111) antiphase domain boundaries. Their calculations showed that the contribution of dislocations and (100) antiphase domain boundaries to the energy was small. Mikkola and Cohen therefore attributed essentially the entire stored energy of shock-loaded Cu_3Au to (111) antiphase boundaries. From the measured values of γ and S, the stored energy could be calculated by eq. (6.26). A comparison with the stored energy values of Beardmore et al.[153] is given in Fig. 67. Mikkola and Cohen did not take into account the stored energy due to disordering of the domains.

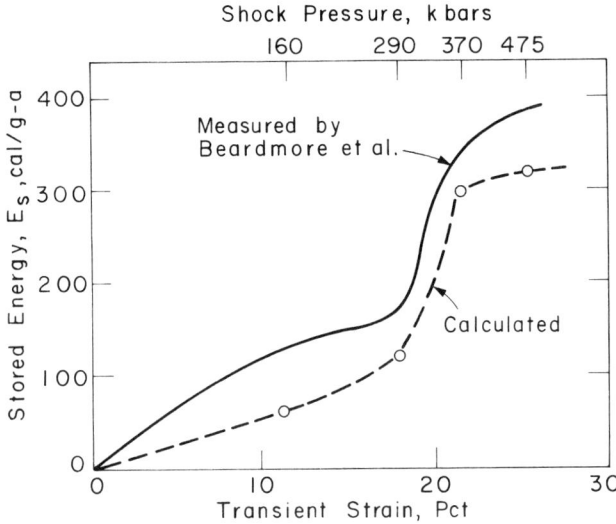

FIG. 67. Comparison of calculated and measured stored energy as a function of strain for ordered Cu_3Au deformed by shock loading. Beardmore et al.,[153] Mikkola and Cohen.[301]

Gangulee and Bever[180] expressed the energy due to the reduction of long-range order as

$$E_s(\text{LRO}) = -\Gamma N_0 \omega [S_0 \exp(-C\bar{\varepsilon})]^2 \quad (6.27)$$

where Γ is a geometrical constant, N_0 is Avogadro's number, S_0 the initial long-range-order parameter and C a constant, and ω is defined as an effective interaction energy. This analysis did not consider the energy associated with antiphase domain boundaries. A separation of the effects of antiphase domain boundaries and reduction of long-range order on the stored energy should be made whenever it is possible.

More attention could well be devoted to the quantitative separation of the stored energy in initially long-range ordered alloys due to the creation of antiphase domain boundaries and disordering from that due to dislocations and point defects. While the latter may be large in absolute magnitude compared to that observed for metallic elements, it is small in comparison with the energy due to the creation of antiphase domain boundaries and the destruction of long-range order.

6.4. KINETICS OF THE RELEASE OF THE STORED ENERGY

The results of investigations of the release of the stored energy were summarized in Part 4; special consideration was given there to the amounts of energy released in each stage of restoration. In the present section various aspects of the release of the stored energy interpreted in section 6.3 will be analyzed in terms of the kinetics of energy release. We shall also discuss the role of the stored energy in recovery and recrystallization.

6.4.1. *Recovery*

Kuhlmann[341] suggested that recovery processes could be described by the relation

$$-\frac{dx}{dt} = cx \exp\left(-\frac{U - mx}{RT}\right) \quad (6.28)$$

where x is the departure at time t of the measured property from equilibrium, U is an activation energy, and c and m are constants. Applied to the recovery process i and the release of the recovery energy E_{pi}, eq. (6.28) becomes

$$-\frac{dE_{pi}}{dt} = c_i E_{pi} \exp\left(-\frac{U_i - m_i E_{pi}}{RT}\right) \quad (6.29)$$

where U_i is the activation energy at the completion of each recovery process and E_{pi} is the instantaneous value of the stored energy associated with the particular process to which the equation applies; $i = 1, 2 \ldots$

Åström[105] considered the following general form of eq. (6.29) in which the several processes i are not distinguished:

$$-\frac{dE_p}{dt} = cE_p \exp\left(-\frac{U - mE_p}{RT}\right) \quad (6.30)$$

According to Åström, if $mE_p \gg RT$, eq. (6.30) reduces to the approximate equation

$$-\frac{dE_p}{dt} = \frac{RT}{m}\frac{1}{t + t'} \quad (6.31)$$

where t' represents the duration of any prior recovery. Borelius et al.[101] found that the rates of energy release during the recovery of aluminum, copper and zinc fitted this equation, but they did not attempt an interpretation.

Åström[105, 106] observed two stages of energy release preceding the recrystallization of 99.98% aluminum deformed by compression. The first stage occurred below 100°C and followed a hyperbolic rate equation of the form of eq. (6.31). The activation energy was 28,000 cal/gram-atom. Åström compared this value with values obtained by Kuhlmann, Masing and Raffelsieper[342] and tentatively identified the process as either climb or, more probably, glide of dislocations. The second stage of energy release, occurring in the temperature range 180° to 250°C, fitted an exponential equation

$$-\frac{dE_p}{dt} = c' \frac{RT}{m\gamma} \exp\left[-c'(t + t')\right] \quad (6.32)$$

where γ is Euler's constant and c' is a constant at a given temperature. This approximate equation follows from eq. (6.30), if $mE_p \ll RT$.[105] For this stage of recovery, Åström obtained an activation energy of 36,000 cal/gram-atom which he took to be the value for self-diffusion in aluminum. He suggested that the process was either the growth of subgrains or polygonization.

Vandermeer and Gordon[172] observed two stages of recovery in aluminum alloyed with 0.0068 at. % copper. The first stage, which occurred at 65°C, followed a hyperbolic rate equation; in the second, which occurred at 139°C, the rate of energy evolution was inversely proportional to the square of the annealing time. Vandermeer and Gordon suggested that the first stage of recovery involved the reduction of the dislocation density within subgrains and rearrangement of dislocations in subgrain walls. They explained the second stage of recovery as slow growth of subgrains subtending small angles of misorientation with their neighbors.

Åström[190] observed a hyperbolic rate of release of energy during the recovery in the range 80° to 186°C of molybdenum deformed by compression. He found an activation energy for the process 1.5 ± 0.2 eV equal to $35,000 \pm 4700$ cal/gram-atom. This activation energy was independent of strain within the accuracy of the data; this suggested that eq. (6.31) might not apply to the recovery of molybdenum. Åström did not put forward possible mechanisms of the recovery process.

Nicholas[304] examined theoretically the kinetics of the release of energy from an initially uniform distribution of point defects migrating to grain boundaries in anisothermal annealing. He compared his results with experimental curves of energy release at three heating rates against temperature for nickel in the range of the Stage IV recovery peak. From the shift in the temperature of the peak with change in heating rate he obtained an activation energy of 1 eV. Curves of energy release against temperature in Stage IV obtained theoretically were in agreement with experimental curves. Nicholas suggested that the recovery was probably due to vacancy migration but could be due to migration of interstitials or vacancy pairs.

Bell and Krisement[109, 110] observed a hyperbolic rate of release of energy during the Stage III recovery of nickel at about 250°C, according to the expression

$$\dot{Q}(t) = \dot{Q}_0 t^{-n} \qquad (6.32a)$$

where $\dot{Q}(t)$ was the isothermal rate of release of energy at time t. They used nickel of two purities (99.6 and 99.9%) and compressed their specimens to 27 and 47% ($\bar{\varepsilon} = -0.32$ and -0.63) at 213°K. The value of n lay between 0.8 and 1.2 and neither n nor $\dot{Q}_0$ depended systematically on the annealing temperature or the purity of the nickel. By carrying out two-stage annealing Bell and Krisement determined an activation energy of 1.1 eV and identified the recovery process as vacancy migration.

Bell[111] investigated the kinetics of Stage III and IV recovery in 99.9% nickel compressed 47% ($\bar{\varepsilon} = -0.63$) at 213°K. Stage III recovery, observed near 100°C, obeyed second-order kinetics and the activation energy was 0.98 eV. Stage IV recovery, which occurred at about 250°C, followed a hyperbolic rate equation. This was a second investigation of this stage of recovery.[109, 110] Bell reported an activation energy of 1.4 eV but noted that at temperatures below about 170°C the value of the activation energy fell to about 1.1 eV.

Gordon, in his investigation of the isothermal release of energy from copper,[104] was unable to determine whether the recovery was characterized by exponential or hyperbolic rate equations.

6.4.1.1. *Energy Release immediately following Deformation*

Williams investigated the kinetics of the release of energy immediately after deformation. With copper tested in the range 18° to 46°C, the kinetic data yielded an activation energy of 14,500 cal/gram-atom.[51] Williams suggested that the release might involve several possible mechanisms including the elimination or trapping of point defects and the rearrangement of dislocations into arrays of lower energy.

In a later investigation[52] of lead, aluminum, silver, nickel, iron and zirconium, Williams noted that for these metals the rate of release immediately after deformation fitted an equation of the form

$$-\frac{dE}{dt} = \frac{1}{kt^2} \qquad (6.33)$$

The kinetics of a second-order process reduce to this equation at long times.[52, 343] Williams considered three mechanisms for the energy release, namely, the motion of dislocations, the elimination of mobile point defects having a spectrum of activation energies for their elimination, and the mutual annihilation or association of point defects. Because the amount of energy released was insensitive to temperature, Williams concluded that the release was probably associated with dislocation movements in all the metals investigated; vacancy effects were likely in copper and silver.

The rate of release from a copper–alluminum and a copper–zinc alloy immediately after deformation[53] did not fit eq. (6.33). Williams believed that the rate resulted from the combined effects of dislocation movements (as in pure metals) and one or more additional processes involving the motion of vacancies, especially vacancy-assisted reordering.

The observations so far considered[51, 53] were made after deformation at high rates. With 99.999% copper extended slowly[54] the rate of release immediately after deformation fitted a hyperbolic equation, a type of equation often observed for conventional recovery.[101, 105, 106, 109, 110, 172] In spite of this difference in kinetics, Williams believed that the mechanism of recovery after slow deformation was essentially the same as that occurring after deformation at high rates.

Ham[39] suggested that the energy release he observed in copper was due to the annealing out of fast-moving defects, perhaps divacancies. On the other hand, he suggested that in a deformed alloy of Cu–3.8Sn vacancies were bound to tin atoms and consequently annealed out slowly causing an energy release that was lost in the exponential cooling curve of the specimen.

As a general point it may be recognized that the substructure of different cold-worked metals, especially if the conditions of cold working differ, are not identical. The kinetics of recovery immediately after deformation, therefore, should not be expected to be the same in all cases. Another factor that may cause differences in these kinetics is the effect of impurities.

6.4.2. *Recrystallization*

Recrystallization is traditionally considered as a nucleation and growth process; the rates of nucleation and growth are usually expressed by equations of the type proposed by Arrhenius. The isothermal rate of growth is assumed to be independent of time. Johnson and Mehl[344] and Avrami[345–347] showed that, if the nucleation rate was also constant, the fraction transformed at constant temperature was given by

$$f = 1 - \exp(-at^n) \tag{6.34}$$

where a and n were constants.

Gordon[104] found that his measurements of the energy evolved during the recrystallization of 99.999% copper fitted such an equation for values of f up to about 0.3 to 0.4. He obtained values of n ranging from 3.1 to 3.9. According to Mehl and Johnson, n should be 4; according to Avrami, it should be between 3 and 4. Gordon obtained an activation energy of 30,400 cal/gram-atom with one lot of copper and 26,400 cal/gram-atom with another lot. He suggested that this difference in the activation energies might be due either to a difference in purity or perhaps in initial grain size. The activation energy of recrystallization appeared to increase with decreasing strain, as is commonly observed.[348] Gordon discussed the possibility that his values of the activation energy might represent the activation energy of grain boundary self-diffusion.

Wenzl[95] investigated the kinetics of the release of energy during the recrystallization of copper of three different purities. He obtained activation energies of recrystallization based on the assumptions that the volume fraction transformed up to a given time was equal to the fraction of the stored energy released, and that the rate of transformation was proportional to the fraction of material untransformed. He also determined the activation energies from the dependence of the recrystallization temperature on the heating rate, following an analysis similar to that of Nicholas.[194] From the first method Wenzl obtained activation energies of 1.0 and 1.7 eV for 99.999 and 99.98% copper, and 2.2 eV for copper containing 0.3% arsenic. From the second he obtained values of 1.2, 1.4 and 2.2 eV. He discussed the relation of these values to the activation energies of vacancy migration and self-diffusion in copper.

From the release of energy by aluminum at 350°C, Åström[105] obtained an activation energy of 58,000 cal/gram-atom which agreed well with the value of 59,000 cal/gram-atom obtained by Anderson and Mehl[348] in a metallographic investigation of the primary recrystallization of 99.95% aluminum. Åström held that his results supported the model developed by Cottrell[349] and Cahn.[350] The essential feature of this model is that the nuclei are pre-existing subgrains which either remained relatively free of strain during deformation or became so during prior recovery.

Vandermeer and Gordon[172] found from measurements of the isothermal release of the stored energy that the kinetics of recrystallization of zone-refined aluminum containing small concentrations of copper followed eq. (6.34) with the exponent n having the value 2 during the early stage of recrystallization. The rate of isothermal recrystallization fell below the predictions of the equation for long times of annealing. This decrease became apparent at shorter times of annealing at lower temperatures or with higher concentrations of copper. Vandermeer and Gordon demonstrated that this retardation was due to the competitive nature of recovery and recrystallization. Under conditions favoring recovery, the stored energy available as a driving force for recrystallization was depleted and the growth rate of recrystallized grains decreased. Vandermeer and Gordon obtained activation energies of 15,000 cal/gram-atom for zone-refined aluminum increasing with copper content to 40,000 cal/gram-atom for aluminum containing 0.0068% copper. They identified these activation energies with those for grain boundary migration.[172]

Bailey[351] and Bailey and Hirsch[108] discussed the recrystallization process in silver, copper and nickel. The copper was from the batch for which Gordon[104] measured the stored energy and determined the kinetics of recrystallization. The release of energy during the recrystallization of silver did not fit the Johnson–Mehl equation with the exception of one batch for which a fit was found over a limited range of time. Observations by electron microscopy indicated that in silver, copper and nickel recrystallization occurred by the migration of original grain boundaries. Such a mechanism has also been observed in aluminum.[298, 299, 352, 353] Bailey[351] proposed a model according to which regions of grain boundary bulged out and migrated under a driving force provided by a difference in dislocation density across the boundary. From this model, Bailey and Hirsch[108] developed a relation from which they were able to obtain values of the exponent n in eq. (6.34) that could be compared with values obtained by applying the Johnson–Mehl equation to the experimental data. The agreement between the values obtained by the two methods was good. Bailey and Hirsch identified the activation energies with those for grain boundary migration. They interpreted the observed variation in the activation energy as due to the effects of impurities and strain on the activation energy for grain boundary migration.

Leighly, Walker and Marx,[354] Walker and Bhattacharya[355] and McNeil and Leighly[356] considered the relation between the stored energy and the recrystallized grain size. They assumed that the stored energy was statistically distributed among regions of the deformed metal that became nuclei of recrystallized grains. It is not clear that such a distribution of the stored energy occurs on this scale, which is the size of one or two dislocation cells.

The interpretations of recrystallization processes discussed in the foregoing have all been based on analyses of isothermal annealing data. Williams[163] discussed the shape of the annealing peaks obtained by Clarebrough et al.[124] for the anisothermal annealing of copper. He suggested that the widths of the peaks were too great to obey annealing kinetics as observed in copper by Gordon.[104] Nicholas[164] pointed out that Williams' analysis was incorrect since the Johnson–Mehl analysis he used applied only to isothermal annealing. Nicholas showed that the proper application of the Johnson–Mehl analysis to anisothermal annealing also predicted that the peak corresponding to recrystallization should be appreciably narrower than the observed peak. He suggested that the greater width of the experimental peak was probably due to the existence of a range of activation energies for recrystallization. He also noted that the variation in the exponent n found by Gordon[104] might in reality be a variation in the distribution of activation energies.

In their investigation of recrystallization of an 82.6Au–17.4Ag alloy after shock loading and after wire drawing, Scattergood et al.[152] found that the activation energies decreased with increasing strain in the wire-drawn specimens and with increasing shock pressure in the shock-loaded specimens. The activation energies of shock-loaded specimens were lower than those of drawn wires with the same amount of stored energy. These authors did not identify the mechanism of recrystallization.

6.4.3. *The Role of the Stored Energy in Restoration Processes*

The stored energy is distinguished from other properties of the cold-worked state in that it is, to a good approximation, the thermodynamic driving force for restoration processes. The stored energy is a small quantity in comparison with the energies of phase transformation; however, it is larger than the total grain boundary energy of a polycrystalline metal of normal grain size. These considerations are important in the application of the formal kinetics of nucleation and growth to recrystallization.

Orowan[357] showed that the relatively small magnitude of stored energy values caused the radius of the critical nucleus and the activation energy of homogeneous nucleation of strain-free grains to be improbably high. Homogeneous nucleation therefore cannot occur. As was pointed out by Christian[358] postulating local regions, such as cell walls in which the stored energy is larger than average, does not alter this conclusion. Heterogeneous nucleation by random thermal fluctuations is also unlikely. It is now generally believed that nuclei pre-exist or are readily formed in cold-worked dislocation structures. Theories of Cahn,[350] Cottrell[349] and Bailey and Hirsch[108] do not require the nucleation of appreciable strain-free regions.

Because the stored energy in extensively cold-worked metals is larger than the total energy of the grain boundaries, the driving force for boundary migration is the stored energy if the difference between the stored energy and the free energy is neglected. Li[343] considered experimental results[359–362] that showed that the growth rate in recrystallization decreased inversely with time at large times; he interpreted these observations as resulting from concurrent recovery which caused the stored energy to be proportional to the inverse of time. The recovery kinetics invoked by Li had been observed by Vandermeer and Gordon.[172]

Dillamore et al.[167] in connection with the development of annealing textures in iron, calculated the stored energy or driving force for recrystallization from measurements of the subgrain boundary density and the misorientation across subgrain boundaries. They were able to show that the observed dependence of these latter quantities on the orientation of the grains in the deformed metal could account for the dominant orientations in the annealing texture.

6.5. The Dissipation of Energy During Deformation

The energy stored during cold working is generally only a small fraction of the expended energy. In this section, we shall briefly consider the mechanisms by which that part of the expended energy that is not stored in the metal is dissipated. This dissipation of energy during deformation constitutes a problem that has received less attention than it deserves.

Nicholas[363] showed that, if the total dissipated energy is to be accounted for, a mobile dislocation is required to dissipate an amount of energy equal to its energy of formation in about 0.1 μm or less of movement. This requirement is quite general and independent of the details of the mechanism by which the moving dislocation dissipates energy.

Nicholas considered four mechanisms for the dissipation of energy by dislocations. These mechanisms are:

(1) the release of the kinetic energy by the arrest of moving dislocations;
(2) thermoelastic damping, radiation damping and scattering of sound waves;
(3) the generation and annihilation of large lengths of dislocation line;
(4) the generation of point defects by moving dislocations and their subsequent annihilation.

Nicholas showed that the first two mechanisms are inadequate unless the speed of the moving dislocations approaches the speed of sound; this speed is not attained.[19] With respect to the third mechanism, Nicholas pointed out that measurements by Seeger et al.[364] of the slip distances in copper indicated that the average distance moved by a dislocation between generation and annihilation was of the order of a few microns or about 10 times the distance within which energy dissipation was required. Thus, the generation and annihilation of dislocations can account at most for about 10% of the dissipated energy. From an analysis of several idealized cases of the fourth mechanism, Nicholas concluded that enough point defects can be generated in a typical metal during cold work at room temperature to account for the dissipated energy provided the point defects can subsequently annihilate each other. Nicholas believed that, in copper at least, the defects would be sufficiently mobile to do so.

Darken[365] discussed the problem of the dissipation of energy during deformation. He considered the frictional loss due to the Peierls–Nabarro force, but concluded that this loss is exceedingly small. He pointed out that the mechanism of energy dissipation was not completely understood and called for further study.

In an experimental investigation, Nakada[68] found that the fraction of the expended energy dissipated during extension was similar for all orientations of single crystals of silver, for all orientations except [100] of single crystals of aluminum and for polycrystalline aluminum. The stress–strain curves of the single crystals exhibited the usual orientation dependence. According to Nakada, analysis of dislocation interactions in crystals having orientations close to the corners of the standard stereographic triangle revealed that tensile deformation of [100] and [110] crystals should produce many more point defects than that of [111] crystals. If the generation and annihilation of point defects had been the dominant mechanism of energy dissipation, [100] and [110] crystals should have shown similar energy dissipation behavior and [111] dissimilar behavior. The energy dissipation behavior of crystals oriented for single slip did not differ from that of other crystals and Nakada suggested that this also tended to rule out the possibility that dissipation occurred by a vacancy mechanism.

Nakada pointed out that the maximum slip distance of approximately 0.1 μm stipulated by Nicholas for dissipation of the self-energy of a dislocation was incorrect if the level of the applied stress was low; at a stress of 3 to 5 kg/mm^2, the distance available for dissipation of the energy was comparable to measured slip distances. Nakada concluded that the generation and annihilation of vacancies might not be an important mechanism of energy dissipation in face-centered cubic metals deformed at room temperature, and that the dominant mechanism might be the generation and annihilation of dislocations.

The preservation of large numbers of point defects after deformation below room temperature (see 6.3.2.1) shows that large numbers can be generated during deformation. The high values of the ratio E_s/E_w at 4.2°K[138, 143a] suggest that the effect of deformation at these temperatures is to retain in the metal a large fraction of the point defects generated during deformation. Since in face-centered cubic metals, the mechanism of deformation is essentially the same at a low temperature as at room temperature, point defects are likely to be generated in comparable numbers at both temperatures. It may be concluded, therefore, in agreement with the conclusion reached by Nicholas, that an appreciable and perhaps the major fraction of the energy dissipated in room-temperature deformation derives from the generation and annihilation of point defects. It is probable that an appreciable amount of energy is also dissipated by annihilation of dislocations, more especially at large strains, but the evidence from stored energy measurements at 4.2°K suggests that, in face-centered cubic metals, this mechanism is subordinate to mechanisms involving point defects.

6.6. Summary of the Contributions of Imperfections to the Stored Energy

The preceding sections of Part 6 have been concerned mainly with the explanation of the effects of experimental variables on the stored energy of cold work. This section will summarize the contributions that different types of imperfections make to the stored energy. It is possible, at least in principle, to allocate to each type of imperfection the amount of stored energy for which it is responsible; some of the required data may be found in section 1.4. Analyses in which the stored energy was partitioned among different imperfections have been published.[130, 148, 180, 301] They are of necessity only approximate, especially because most do not take account of types of possible interaction between different imperfections. These analyses, however, give a useful picture of the relative importance of different mechanisms for storing energy.

6.6.1 Dislocations

Dislocations are a type of imperfection always present in cold-worked metals. They account for the energy released during recrystallization. They probably also contribute an appreciable part of the energy released in any recovery stage immediately preceding recrystallization. Deformation at low temperatures causes a higher density of dislocations or dislocation arrangements of higher energy or both than deformation at room temperature. The relative importance of dislocations as a mechanism of storing energy, however, decreases as the temperature of deformation is lowered. Of the energy stored at 4.2°K, only a small fraction additional to that stored at 77°K appears to be due to dislocations.

The absolute magnitude of the contribution of dislocations to the stored energy increases as the concentration of impurities in a metal increases. Most of this additional energy is probably due to the pinning of dislocations by impurity atoms and is released during recovery immediately preceding recrystallization. In alloys the fraction of the stored energy due to dislocations is in general appreciably smaller than in pure metals.

6.6.2. Point Defects

Point defects contribute only minor amounts to the stored energy at temperatures of deformation above a certain fraction of the absolute temperature of recrystallization T_r, namely, approximately $2/3\ T_r$. As the temperature of deformation decreases, however, the contribution of point defects increases, and at and below 77°K, appreciable amounts of energy are stored in a typical metal in the form of point defects. The precise arrangement of point defects present after any given type of cold work and the order in which they anneal out remain uncertain at present. There is little doubt, however, that most, and in some cases all, of the energy released by annealing below room temperature is due to the elimination of point defects. In metals having a high recrystallization temperature, some recovery processes occurring above room temperature are also due to point defects. In addition, point defects may play a role in the recovery immediately preceding recrystallization in relatively impure metals. They appear to be important in promoting reordering in alloys disordered by deformation.

As noted in section 6.5, the creation and disappearance of point defects probably constitute the principal mechanism of the dissipation of energy during deformation.

6.6.3. Stacking Faults—Twins and Twin Faults

Stacking faults in cold-worked metals may be considered as inherent features of extended dislocations or as a separate species of imperfection. If faults are present in large numbers, the faulted areas may make a small contribution to the stored energy. Their most important role, however, may be in governing the occurrence of cross slip and hence the extent of dislocation annihilation taking place during deformation. This may affect the amounts of energy stored by different metals.

Deformation twins on a scale visible under the optical microscope cannot account for an appreciable amount of stored energy. Obviously, their contribution would be smaller than the change in energy resulting from the increase in grain boundary area when the grain size is reduced. Twinning on a submicroscopic scale (microtwinning or twin-faulting) may make a small direct contribution to the stored energy. This may be of approximately the same magnitude as the contributions of stacking faults. An indirect and possibly more important contribution may result from changes in the dislocation arrangement accompanying the onset of microtwinning, in particular, the occurrence of a large dislocation density at the twin interfaces. In face-centred cubic metals, however, deformation twinning, including microtwinning, occurs only at low temperatures or high rates of deformation.

6.6.4. *Destruction of Order*

In solid solution alloys an important general mechanism of energy storage is the mechanical destruction of short-range order or clustering. The relative importance of this mechanism depends on the particular alloy and the degree of order it can develop. It may contribute the major fraction of the stored energy in some alloys. The energy due to the destruction of order supplements that due to the mechanisms of storage involving dislocations and point defects, which are always operative: this is one factor causing the energy stored by an alloy to be appreciably larger than that stored in the component metals. Another factor is the retarding effect of a solute on recovery processes.

In alloys in which long-range order develops, the stored energy can be especially large, since deformation creates antiphase domain boundaries and also reduces the long-range order within the domains. The mechanically and thermally disordered states are not identical. Alloys with long-range order which have been thermally disordered retain some degree of short-range order. They can therefore store energy by the mechanical destruction of short-range order. The amount of energy stored in this way is comparable to the amount of energy stored by the same mechanism in alloys that do not develop long-range order.

The deformation of an intermetallic compound reduces its order in the same manner as the reduction of order by the deformation of an ordered solid solution alloy. The magnitude of the stored energy in a cold-worked compound may be expected to be related to the magnitude of its heat of formation.

CONCLUSION

Many aspects of the stored energy of cold work have become understood in recent years, but other problems still remain to be clarified and new ones have appeared. The outstanding problem is the large spread in the values of the stored energy measured in different laboratories with nominally the same material, specifically copper. In this review we have suggested explanations for the spread: they involve the identity, amount and distribution of impurities and the method of measurement. Work on specimens of identical material in different laboratories would be one way to establish firmly the merits of the different methods for any given type of stored energy measurement; this should produce values free from ambiguity, thus providing a firm base for interpretation of their meaning rather than their magnitudes. Such cooperative measurements have never been carried out.

The manner in which certain variables affect the stored energy is established. This is so for the extent of deformation, the temperature, the grain size and the identity of the metal. The effects of other variables call for further investigation. The deformation process appears to affect the saturation level of the stored energy, where such saturation occurs. Differences in the deformation path and in the temperature rise during cold working may account for the differences in saturation levels, but further investigation of the problem is desirable. The effect of impurities on the amount of the stored energy has not been investigated systematically. With the techniques of material preparation and analysis now available, experiments in which the levels, types and distribution of impurities are controlled have become possible.

Various relations between structural features and the stored energy still need to be explored. With respect to crystal structure more work on body-centered cubic metals and particularly on non-cubic metals should be carried out. Also systematic investigations of the energy stored by single crystals, which have been started only recently, should be extended. Preferred orientation and pre-existing substructure are other structural features that deserve investigation. To date the energy stored by multiphase alloys has received little attention. Dispersion-hardened alloys are of special interest in this connection. The energy stored by heat-treated alloys such as age-hardened alloys should be investigated. Solid solutions that exhibit clustering pose an especially interesting problem.

The various mechanisms of storing energy are not fully understood. The contribution of dislocations to the stored energy can be estimated with fair accuracy if the dislocation density is known. The dependence of the dislocation density on variables related to the deformation process and on variables related to the metal, however, cannot generally be predicted. The contribution of point defects to the stored energy can be estimated provided their type and number are known. The mechanisms for producing point defects are conjectural, the type and number present involve a running balance between the generation and annealing-out of these defects, and the prediction of their type and number is not possible at present. The frequency of stacking faults and twin faults in a deformed metal is also not predictable. The energy due to the destruction of long-range or short-range order involves both the relation between the energy content and the degree of order, and the effects of the deformation variables on the degree of order. For long-range ordered alloys it is uncertain how much energy is to be associated with antiphase domain boundaries and how much with destruction of order within domains.

In summary, the relation of the stored energy to crystal imperfections is better understood than is the relation of crystal imperfections to variables involving the metal and the deformation process. The ultimate aim of theoretical analysis must be the development of a detailed model of the cold-worked structure resulting from the deformation process. The stored energy, as a measure of the integrated effects of imperfections, will then be predictable from theory.

ACKNOWLEDGMENTS

We are indebted to various persons and institutions for assistance and support in the writing of this review. We have benefited from discussions and correspondence, particularly with Drs. L. M. Clarebrough, J. L. White, R. O. Williams and A. Wolfenden, and Professors J. B. Cohen and M. E. Hargreaves. We thank Professors C. Wagner, R. Schuhmann Jr. and J. R. Rice for discussions of questions involving thermodynamics.

M.B.B. is grateful for the opportunity to study fundamental aspects of the stored energy of cold work during a sabbatical year spent at Harvard University as a guest of the Division of Engineering and Applied Physics. D. L. H. acknowledges financial support by the Office of Naval Research during part of the period in which this review was written. A. L. T. thanks the University of Auckland for sabbatical leave, which made our collaboration possible. General support of the work on this review and of related experimental research by the U.S. Atomic Energy Commission is also acknowledged.

REFERENCES

1. TITCHENER, A. L. and BEVER, M. B. *Progress in Metal Physics* **7** (1958) 247.
2. BEVER, M. B. In *Creep and Recovery*, A.S.M. (1957) 14.
3. CLAREBROUGH, L. M., HARGREAVES, M. E. and LORETTO, M. H. In *Recovery and Recrystallization of Metals*, ed. by L. Himmel, Interscience (1963) 63.
4. BEVER, M. B. In *Symposium on the Role of Substructure in the Mechanical Behaviour of Metals*, Orlando, Florida, December 1962; Technical Report No. ASD–TDR–63–324, April 1963, Directorate of Materials and Processes, Air Force Systems Command, Wright–Patterson Air Force Base, Ohio.
5. WILLIAMS, R. O. In *Experimental Methods of Materials Research*, ed. by H. Herman, Interscience, Vol. 1 (1967) 251.
6. GUBKIN, S. I. and BOGDANOV, YE. S. *Dokl. Akad. Nauk SSSR* **88** (1953) 967.
7. THOMSON, W. (Lord Kelvin) *Trans. Roy. Soc. Edinb.* **20** (1853) 261; *Math. and Phys. Papers*, Cambridge, Vol. 1 (1882) 174.
8. THOMSON, W. (Lord Kelvin) *Quart. J. of Pure and Appl. Math.* **1** (1857) 57; *Math. and Phys. Papers*, Cambridge, Vol. 1 (1882) 291.
9. ROCCA, R. and BEVER, M. B. *Trans. Amer. Inst. Min. (Metall.) Engrs.* **188** (1950) 327.
10. TAMMANN, G. and WARRENTRUP, H. *Z. Metallk.* **29** (1937) 84.
11. BRIDGMAN, P. W. *Rev. Mod. Phys.* **22** (1950) 56.
12. REGNAULT, V. *Annal. Phys. Chem.* **51** (1840) 213.
13. EUCKEN, A. In *Handbuch der Experimentalphysik*, Akademische Verlags-Gesellschaft, Vol. 8, Pt. 1 (1929) 195.
14. SUZUKI, T. *Sci. Rep. Tôhoku Univ.* **A1** (1949) 193.
15. MAIER, C. G. and ANDERSON, C. T. *J. Chem. Phys.* **2** (1934) 513.
16. MARTIN, D. L. *Canad. J. Phys.* **38** (1960) 17.
17. AHLERS, G. *Rev. Sci. Instr.* **37** (1966) 477.
18. PERVAKOV, V. A. and KHOTKEVICH, V. I. *Ukrain. Fiz. Zhur.* **12** (1967) 1777. (Eng. Trans. p. 1697).
19. COTTRELL, A. H. *Dislocations and Plastic Flow in Crystals*, Oxford (1953).
20. FRIEDEL, J. *Dislocations*, Pergamon (1964).
21. GEHLEN, P. C. *J. Appl. Phys.* **41** (1970) 5165.
22. SEEGER, A. In *Handbuch der Physik*, Springer, Vol. 7, Pt. 1 (1955) 546.
23. LI, J. C. M. *J. Appl. Phys.* **32** (1961) 1873.
24. VINEYARD, G. H. and DIENES, G. J. *Phys. Rev.* **93** (1954) 265.
25. SIMMONS, R. O. and BALLUFFI, R. W. *Phys. Rev.* **129** (1963) 1533.
26. PRIGOGINE, I. *Introduction to Thermodynamics of Irreversible Processes*, Wiley (1967).
27. DE GROOT, S. R. and MAZUR, P. *Non-Equilibrium Thermodynamics*, North Holland (1962).
28. KESTIN, J. In *Irreversible Aspects of Continuum Mechanics and Transfer of Physical Characteristics in Moving Fluids*, ed. by H. Parkus and L. I. Sedov, Springer (1968) 117.
29. RUBIN, D. *Trans. ASME, J. Appl. Mech.* **35** (1968) 596.
30. MEIXNER, J. and REIK, H. G. In *Encyclopedia of Physics*, ed. by S. Flügge, Springer, Vol. 3, Pt. 2 (1959) 413.
31. MEIXNER, J. *Ned. Tijdschrift Natuurkunde* **26**, No. 9 (1960) 259.
32. COLEMAN, B. D. and GURTIN, M. E. *J. Chem. Phys.* **47** (1967) 597.
33. KESTIN, J. and RICE, J. R. In *A Critical Review of Thermodynamics*, ed. by E. B. Stuart, B. Gal-Or and A. J. Brainard, Mono Book (1970) 275.

34. CHU, BOA-TEH. In *A Critical Review of Thermodynamics,* ed. by E. B. Stuart, B. Gal-Or and A. J. Brainard, Mono Book (1970) 299.
35. KRATOCHVIL, J. and DILLON. O. W., JR. *J. Appl. Phys.* **40** (1969) 3207.
36. HILL, R. *The Mathematical Theory of Plasticity,* Oxford (1956).
36a. RICE, J. R. *Trans. ASME, J. Appl. Mech.* **37** (1970) 728.
36b. RICE, J. R. *J. Mech. Phys. Solids.* **19** (1971) 433.
37. LEACH, J. S. LL. In *Physicochemical Measurements in Metals Research,* ed. by R. A. Rapp, Interscience, Vol. 4, Pt. 1 (1970) 197.
38. WILLIAMS, R. O. *Rev. Sci. Instr.* **31** (1960) 1336.
39. HAM, R. K. *Phil. Mag.* **15** (1967) 257.
40. TAYLOR, G. I. and QUINNEY, H. *Proc. Roy. Soc.* **A143** (1934) 307.
41. ROSENHAIN, W. and STOTT, V. H. *Proc. Roy. Soc.* **A140** (1933) 9.
42. EPIFANOV, G. I. and REBINDER, P. A. *Dokl. Akad. Nauk. SSSR* **66** (1949) 653.
43. WILLIAMS, R. O. *Rev. Sci. Instr.* **34** (1963) 639.
44. ERDMANN, J. C. and JAHODA, J. A. *Rev. Sci. Instr.* **34** (1963) 172.
45. WOLFENDEN, A. and APPLETON, A. S. *Rev. Sci. Instr.* **38** (1967) 826.
46. CLAREBROUGH, L. M., HARGREAVES, M. E. and WEST, G. W. *Proc. Roy. Soc.* **A232** (1955) 252.
47. HORT, H. *Z. Ver. Dtsch. Ing.* **50** (1906) 1831.
48. HORT, H. *Mitt. ForschArb. Ingenieurw.* **41** (1907) 1.
49. CHARBONNIER, P. and GALY-ACHÉ. *Mémorial de l'Artillerie de la Marine* **28** (1900) 391.
50. WILLIAMS, R. O. *Acta Met.* **12** (1964) 745.
51. WILLIAMS, R. O. *Acta Met.* **9** (1961) 949.
52. WILLIAMS, R. O. *Trans. Met. Soc. AIME* **224** (1962) 719.
53. WILLIAMS, R. O. *Trans. Met. Soc. AIME* **227** (1963) 1290.
54. WILLIAMS, R. O. *Acta Met.* **13** (1965) 163.
55. WOLFENDEN, A. *Acta Met.* **15** (1967) 971.
56. WOLFENDEN, A. *Scripta Met.* **4** (1970) 327.
57. WOLFENDEN, A. *Acta Met.* **17** (1969) 585.
58. WOLFENDEN, A. *Scripta Met.* **2** (1968) 621.
59. WOLFENDEN, A. *Scripta Met.* **3** (1969) 429.
60. WOLFENDEN, A. *Scripta Met.* **3** (1969) 709.
61. DILLON, O. W., JR. *J. Mech. Phys. Solids* **10** (1962) 235.
62. DILLON, O. W., JR. *J. Appl. Phys.* **33** (1962) 3100.
63. DILLON, O. W., JR. *J. Mech. Phys. Solids* **11** (1963) 21.
64. DILLON, O. W., JR. *Int. J. Solids Structures* **2** (1966) 181.
65. KUNIN, V. N. *Fiz. Met. i Metalloved.* **7** (1959) 790; *Phys. Metals and Metallog.* **7** (5) (1959) 136.
66. KUNIN, V. N. *Fiz. Met. i Metalloved.* **8** (1959) 17; *Phys. Metals and Metallog.* **8** (1) (1959) 16.
67. KUNIN, N. F., KUNIN, V. N., GRISHKEVICH, A. YE. and KORENCHENKO, YE. S. *Fiz. Met. i Metalloved.* **17** (1964) 789; *Phys. Metals and Metallog.* **7** (5) (1964) 150.
68. NAKADA, Y. *Phil. Mag.* **11** (1965) 251.
69. HALFORD, G. R. *Stored Energy of Cold Work Changes Induced by Cyclic Deformation,* Ph.D. Thesis, University of Illinois, Urbana (1966).
70. PANIN, V. E. and MILEVSKAYA, V. G. *Fiz. Met. i Metalloved.* **5** (1957) 120; *Phys. Metals and Metallog.* **5** (1) (1957) 97.
71. PANIN, V. E., SUKHOVAROV, E. and DUDAREV, E. F. *Izvest. vusov Fiz.* **8** (1967) 152.
72. SHERMERGOR, T. D. *Fiz. Met. i Metalloved.* **7** (1959) 146; *Phys. Metals and Metallog.* **7** (1) (1959) 139.
73. SATO, S. *Sci. Rep. Tôhoku Univ.* **20** (1931) 140.
74. WHITE, J. L. and KOYAMA, K. *Rev. Sci. Instr.* **34** (1963) 1104.
75. CLAREBROUGH, L. M., HARGREAVES, M. E., MICHELL, D. and WEST, G. W. *Proc. Roy. Soc.* **A215** (1952) 507.
76. KANZAKI, H. *J. Phys. Soc. Japan* **6** (1951) 90.
77. QUINNEY, H. and TAYLOR, G. I. *Proc. Roy. Soc.* **A163** (1937) 157.
78. WELBER, B. *J. Appl. Phys.* **23** (1952) 876.
79. KANZAKI, H. *J. Phys. Soc. Japan* **6** (1951) 456.
80. WELBER, B. and WEBELER, R. *Trans. Amer. Inst. Min. (Metall.) Engrs.* **197** (1953) 1558.
81. MIMA, G. and TOKIZAWA, M. *Nippon Kinzoku Gakkai-Si* **23**, No. 5 (1959) 307.
82. POPOV, M. M., TIKHOMINA, E. N., SKURATOV, S. M. and KALININA, E. N. *Fiz. Met i Metalloved.* **8** (1959) 103; *Phys. Metals and Metallog.* **8** (1) (1959) 87.
83. KOVACS, I. In *Ber. Internat. Symposium Reinstoffe in Wissenschaft u. Technik,* Dresden, 1965; Teil 3: *Realstruktur u. Eigenschaften von Reinstoffen* (1967) 719.
84. CHIN, L. L. J. and GRANT, N. J. *Powder Metallurgy* **10** (1967) 344.
85. CHIN, L. L. J. and GRANT, N. J. In *Proc. Second Bolton Landing Conf. on Oxide Dispersion Strengthening,* ed. by G. S. Ansell, T. D. Cooper and F. V. Lenel, Gordon & Breach (1968) 213.

86. MISRA, S., HOWLETT, B. W. and BEVER, M. B. *Trans. Met. Soc. AIME* **233** (1965) 749.
87. KHOTKEVICH, V. I., CHAIKOVSKII, E. F. and ZASHKVARA, V. V. *Dokl. Akad. Nauk SSSR* **96** (1954) 483.
88. HENDERSON, J. W. and KOEHLER, J. S. *Phys. Rev.* **104** (1956) 626.
89. VAN DEN BEUKEL, A. *Physica* **27** (1961) 603.
90. LORETTO, M. H., HARGREAVES, M. E. and CLAREBROUGH, L. M. *J. Aust. Inst. Metals* **8** (1963) 127.
91. VAN DEN BEUKEL, A. *Acta Met.* **11** (1963) 97.
92. LUGSCHEIDER, W. and WILDHACK, H. *Z. Metallk.* **59** (1968) 124.
93. KRIVOBOK, V. N. *Trans. Amer. Soc. Steel Treat.* **8** (1925) 703.
94. SIZMANN, R. and WENZL, H. *Z. Angew. Phys.* **11** (1959) 362.
94a. WENZL, H. *Z. Angew. Phys.* **15** (1963) 172.
95. WENZL, H. *Z. Angew. Phys.* **15** (1963) 286.
96. HERTSRIKEN, S. D., LARIKOV, L. N. and SLYUSAR, B. P. *Ukrain. Fiz. Zhur.* **5** (1960) 672.
97. TAOKA, T., SUZUKI, K., YOSHIKAWA, A. and OKAMOTO, M. *Acta Met.* **13** (1965) 1311.
98. LUGSCHEIDER, W. *Ber. Bunsenges. Phys. Chem.* **71** (1967) 228.
99. EBEL, H. and LUGSCHEIDER, W. *Z. Metallk.* **57** (1966) 363.
100. BORELIUS, G. In *L'Etat Solide*, Institut International de Physique Solvay, Neuviéme Conseil de Physique (1952) 427.
101. BORELIUS, G., BERGLUND, S. and SJÖBERG, S. *Ark. Fys.* **6** (1952) 143.
102. BORELIUS, G., BERGLUND, S. and AVSAN, O. *Ark. Fys.* **2** (1951) 551.
103. BORELIUS, G. and BERGLUND, S. *Ark. Fys.* **4** (1951) 173.
104. GORDON, P. *Trans. Amer. Inst. Min. (Metall.) Engrs* **203** (1955) 1043.
105. ÅSTRÖM, H. U. *Ark. Fys.* **10** (1955) 197.
106. ÅSTRÖM, H. U. *Acta Met.* **3** (1955) 508.
107. BAILEY, J. E. and HIRSCH, P. B. *Phil. Mag.* **5** (1960) 485.
108. BAILEY, J. E. and HIRSCH, P. B. *Proc. Roy. Soc.* **A267** (1962) 11.
109. BELL, F. and KRISEMENT, O. *Z. Metallk.* **53** (1962) 115.
110. BELL, F. and KRISEMENT, O. *Acta Met.* **10** (1962) 80.
111. BELL, F. *Acta Met.* **13** (1965) 363.
112. BELL, F. *Arch. Eisenhüttenwesen* **36** (1965) 745.
113. SMITH, C. J. *Proc. Roy. Soc.* **A125** (1929) 619.
114. FRANCE, R. W. *Trans. Faraday Soc.* **30** (1934) 450.
115. KOREF, F. and WOLFF, H. *Z. Elektrochem.* **28** (1922) 477.
116. VAN LIEMPT, J. A. M. *Z. Anorg. Chem.* **129** (1923) 263.
117. TICKNOR, L. B. and BEVER, M. B. *Trans. Amer. Inst. Min. (Metall.) Engrs* **194** (1952) 941.
118. HOWLETT, B. W., LEACH, J. S. LL., TICKNOR, L. B. and BEVER, M. B. *Rev. Sci. Instr.* **33** (1962) 619.
119. KUNIN, N. F. and SENILOV, G. V. *Trud. Sib. Fiz. Tekh. Inst.* **4** (3) (1936) 132.
120. FEDOROV, A. A. *J. Tech. Phys. Moscow* **11** (1941) 999.
121. TIZHNOVA, N. V. *J. Tech. Phys. Moscow* **16** (1946) 1389.
122. STUDENOK, IU. A. *J. Tech. Phys. Moscow* **20** (1950) 431.
123. DEGTIAREV, M. M. *J. Tech. Phys. Moscow* **20** (1950) 440.
124. CLAREBROUGH, L. M., HARGREAVES, M. E. and LORETTO, M. H. *Acta Met.* **6** (1958) 725.
125. WHITE, J. L. In *Recovery and Recrystallization of Metals*, ed. by L. Himmel, Interscience (1963) 122.
126. BEVER, M. B. and TICKNOR, L. B. *Acta Met.* **1** (1953) 116.
127. BEVER, M. B., MARSHALL, E. R. and TICKNOR, L. B. *J. Appl. Phys.* **24** (1953) 1176.
128. TITCHENER, A. L. and BEVER, M. B. *Trans. Met. Soc. AIME* **215** (1959) 326.
129. SMITH, J. H. and BEVER, M. B. *Trans. Met. Soc. AIME* **242** (1968) 880.
130. COHEN, J. B. and BEVER, M. B. *Trans. Met. Soc. AIME* **218** (1960) 155.
131. WOLFENDEN, A. *Acta Met.* **19** (1971) 1373.
132. BEVER, M. B. and TICKNOR, L. B. *J. Appl. Phys.* **22** (1951) 1297.
133. MASIMA, M. and SACHS, G. *Z. Phys.* **56** (1929) 394.
134. LEACH, J. S. LL., LOEWEN, E. G. and BEVER, M. B. *J. Appl. Phys.* **26** (1955) 728.
135. GREENFIELD, P. and BEVER, M. B. *Acta Met.* **4** (1956) 433.
136. GREENFIELD, P. and BEVER, M. B. *Acta Met.* **5** (1957) 125.
137. ROESSLER, B. and BEVER, M. B. *Trans. Met. Soc. AIME* **221** (1961) 1049.
138. APPLETON, A. S. and BEVER, M. B. *Trans. Met. Soc. AIME* **227** (1963) 365.
139. BOGACHEV, I. N. and DENISOVA, I. K. *Fiz. Met. i Metalloved.* **28** (1969) 1118; *Phys. Metals and Metallog.* **28** (6) (1969) 177.
140. ROBINSON, P. M. and BEVER, M. B. *Acta Met.* **13** (1965) 647.
141. ———
142. RIGGS, F. B. JR. Calorimetry of Deformed Copper. Doct. Phil., Harv. (1956).
143. WANG, T. P. and BROWN, N. *Trans. Amer. Soc. Metals* **50** (1958) 541.
143a. ERDMANN, J. C. and JAHODA, J. A. *Appl. Phys. Letters* **4** (1964) 204.

143b. ERDMANN, J. C. and JAHODA, J. A. In *Progress Review, Solid State Physics Laboratory*. Boeing Scientific Research Laboratories (Feb. 1963) 121.
144. ERDMANN, J. C. and JAHODA, J. A. In *Review, Solid State Physics Laboratory July–December 1964*, Boeing Scientific Research Laboratories (Feb. 1965) 100.
145. ERDMANN, J. C. and JAHODA, J. A. *Appl. Phys. Letters* **13** (1968) 393.
146. WOLFENDEN, A. and APPLETON, A. S. *Acta Met.* **16** (1968) 915.
147. WOLFENDEN, A. *Acta. Met.* **16** (1968) 975.
148. AVERBACH, B. L., BEVER, M. B., COMERFORD, M. F. and LEACH, J. S. LL. *Acta Met.* **4** (1956) 477.
149. MICHELL, D. and HAIG, F. D. *Phil. Mag.* **2** (1957) 15.
150. APPLETON, A. S., DIETER, G. E. and BEVER, M. B. *Trans. Met. Soc. AIME* **221** (1961) 90.
151. IYER, A. S. and GORDON, P. *Trans. Met. Soc. AIME* **224** (1962) 1077.
152. SCATTERGOOD, R. O., BEARDMORE. P. and BEVER, M. B. *Trans. Met. Soc. AIME* **227** (1963) 1468.
153. BEARDMORE, P., HOLTZMAN. A. H. and BEVER, M. B. *Trans. Met. Soc. AIME* **230** (1964) 725.
154. BRILLHART, D. C., DE ANGELIS, R. J., PREBAN, A. G., COHEN, J. B. and GORDON, P. *Trans. Met. Soc. AIME* **239** (1967) 836.
155. GORDON, P. *Rev. Sci. Instr.* **25** (1954) 1173.
156. POLAKOWSKI, N. H. *J. Iron Steel Inst.* **185** (1957) 67.
157. IYER, A. S. and GORDON, P. *Trans. Met. Soc. AIME* **215** (1959) 729.
158. HARGREAVES, M. E., LORETTO, M. H., CLAREBROUGH, L. M. and SEGALL, R. L. In *N.P.L. Symp. No. 15, Teddington 1963*, H.M.S.O. (1963) 209.
159. TITCHENER, A. L. and BEVER, M. B. *Acta Met.* **8** (1960) 338.
160. CLAREBROUGH, L. M., HARGREAVES, M. E., HEAD, A. K. and WEST, G. W. *Trans. Amer. Inst. Min. (Metall.) Engrs* **203** (1955) 99.
161. CLAREBROUGH, L. M., HARGREAVES, M. E., WEST, G. W. and HEAD, A. K. *Proc. Roy. Soc.* **A242** (1957) 160
162. DILLON, O. W., JR. In *Mechanical Behavior of Materials under Dynamic Loads*, ed. by U. S. Lindholm, Springer-Verlag (1968) 21.
163. WILLIAMS, R. O. *Acta Met.* **7** (1959) 676.
164. CLAREBROUGH, L. M., HARGREAVES, M. E. and LORETTO, M. H. (Pt. 1), and NICHOLAS, J. F. (Pt. 2) *Acta Met.* **8** (1960) 736.
165. TITCHENER, A. L. *Acta Met.* **9** (1961) 379.
166. LORETTO, M. H. and WHITE, A. J. *Acta Met.* **9** (1961) 512.
167. DILLAMORE, I. L., SMITH, C. J. E. and WATSON, T. W. *Met. Sci. J.* **1** (1967) 49.
168. RICHARDS, C. E. and WATSON T. W. *J. Iron Steel Inst.* **207** (1969) 582.
169. FARREN, W. S. and TAYLOR, G. I. *Proc. Roy. Soc.* **A107** (1925) 422.
170. CLAREBROUGH, L. M., HARGREAVES, M. E., LORETTO, M. H. and WEST, G. W. *Acta Met.* **8** (1960) 797.
171. MICHELL, D. and LOVEGROVE, E. *Phil. Mag.* **5** (1960) 499.
172. VANDERMEER, R. A. and GORDON, P. In *Recovery and Recrystallization of Metals*, ed. by L. Himmel. Interscience (1963) 211.
173. BOHNENKAMP, K., LÜCKE, K. and MASING, G. *Z. Metallk.* **46** (1955) 765.
174. CLAREBROUGH, L. M., HARGREAVES, M. E. and WEST, G. W. *Acta Met.* **5** (1957) 738.
175. CLAREBROUGH, L. M., HARGREAVES, M. E. and LORETTO, M. H. *Phil. Mag.* **6** (1961) 807.
176. CLAREBROUGH, L. M., HARGREAVES, M. E. and LORETTO, M. H. *Phil. Mag.* **7** (1962) 115.
177. SCHOTTKY, W. F. and BEVER, M. B. *Acta Met.* **7** (1959) 199.
178. KHOTKEVICH, V. I. and SIRENKO, G. A. *Ukrain. Fiz. Zhur.* **14** (1969) 1558.
179. CLAREBROUGH, L. M., HARGREAVES, M. E. and LORETTO, M. H. *Proc. Roy. Soc.* **A257** (1960) 363.
180. GANGULEE, A. and BEVER, M. B. *Phil. Mag.* **20** (1969) 519.
181. WALDMAN, J. and BEVER, M. B. Unpublished work.
182. WOLFENDEN, A. *Scripta Met.* **5** (1971) 371.
183. ROBINSON, P. M. and BEVER, M. B. *Acta Met.* **14** (1966) 693.
184. MEECHAN, C. J. and BRINKMAN, J. A. *Phys. Rev.* **103** (1956) 1193.
185. DIENES, G. J. and VINEYARD, G. H. *Radiation Effects in Solids*, Interscience (1957).
186. BALLUFFI, R. W., KOEHLER, J. S. and SIMMONS, R. O. In *Recovery and Recrystallization of Metals*, ed. by L. Himmel, Interscience (1963) 1.
187. CHUANG, K. C. and BEVER, M. B. Unpublished work.
188. MICHELL, D. *Phil. Mag.* **1** (1956) 584.
189. CLAREBROUGH, L. M., HARGREAVES, M. E. and WEST, G. W. *Phil. Mag.* **44** (1953) 913.
190. ÅSTRÖM, H. U. *Ark. Fys.* **26** (1963) 83.
191. CLAREBROUGH, L. M. and LORETTO, M. H. *Proc. Roy. Soc.* **A257** (1960) 326.
192. CLAREBROUGH, L. M., HARGREAVES, M. E. and LORETTO, M. H. *Proc. Roy. Soc.* **A257** (1960) 338.
193. CLAREBROUGH, L. M., HARGREAVES, M. E. and LORETTO, M. H. *Proc. Roy. Soc.* **A261** (1961) 500.
194. NICHOLAS, J. F. *Aust. J. Phys.* **9** (1956) 425.

195. PERVAKOV, V. A., KHOTKEVICH, V. I. and SHEPELEV, A. G. *Fiz. Met. i Metalloved.* **10** (1960) 117; *Phys. Metals and Metallog.* **10** (1) (1960) 107.
196. BEARDMORE, P. and BEVER, M. B. *Trans. Met. Soc. AIME* **245** (1969) 165.
197. CLAREBROUGH, L. M., HARGREAVES, M. E. and WEST, G. W. *Phil. Mag.* **1** (1956) 528.
198. CLAREBROUGH, L. M., HARGREAVES, M. E. and LORETTO, M. H. *J. Aust. Inst. Metals* **6** (1961) 104.
199. FASTOV, N. S. *Prob. Metalloved. i Fiziki Metallov.*, 4th Symp., Gosu Nanchno-Tekhn. Liter. Izdat po Chernoii Tsvetnoi Metallurgii. Moscow (1955) 377.
200. BAILEY, J. E. *Phil. Mag.* **8** (1963) 223.
201. HIRN, G. A. *Théorie Mécanique de la Chaleur*, Gauthiers-Villars, Vol. 1 (1875) 95.
202. ROSENHAIN, W. In *Dictionary of Applied Physics*. Macmillan, Vol. 5 (1923) 398.
203. RUSSELL, T. F. *J. Iron Steel Inst.* **57** (1925) 497.
204. MAIER, C. G. *Trans. Amer. Inst. Min. (Metall.) Engrs* **122** (1936) 121.
205. MCADAM, D. J., JR. *Trans. Amer. Soc. Metals* **43** (1951) 970.
206. MCADAM, D. J., JR. *Trans. Amer. Soc. Metals* **43** (1951) 1215.
207. MCADAM, D. J., JR. *Trans. Amer. Inst. Min. (Metall.) Engrs* **185** (1949) 727.
208. GEISS, W. and VAN LIEMPT, J. A. M. *Z. Anorg. Chem.* **143** (1925) 259.
209. GEISS, W. and VAN LIEMPT, J. A. M. *Z. Phys.* **41** (1927) 867.
210. VAN LIEMPT, J. A. M. *Z. Anorg. Chem.* **195** (1931) 366.
211. WOOD, W. A. *Phil. Mag.* **18** (1934) 495.
212. BOAS, W. *Z. Kristallogr.* **96** (1937) 214.
213. BOAS, W. *Z. Kristallogr.* **97** (1937) 354.
213a. TAYLOR, G. I. *The Scientific Papers of Sir Geoffrey Ingram Taylor*, ed. by G. K. Batchelor, Cambridge, Vol. 1 (1958) 399.
214. FAULKNER, E. A. *Phil. Mag.* **5** (1960) 843.
215. FAULKNER, E. A. and HAM, R. K. *Phil. Mag.* **7** (1962) 279.
216. HAWORTH, F. E. *Phys. Rev.* **52** (1937) 613.
217. STIBITZ, G. R. *Phys. Rev.* **49** (1936) 862.
218. CAGLIOTI, V. and SACHS, G. *Z. Phys.* **74** (1932) 647.
219. BRINDLEY, G. W. and RIDLEY, P. *Proc. Phys. Soc. Lond.* **51** (1939) 432.
220. WOOD, W. A. *Proc. Roy. Soc.* **A172** (1939) 231.
221. SMITH, C. S. and STICKLEY, E. E. *Phys. Rev.* **64** (1943) 191.
222. STOKES, A. R., PASCOE, K. J. and LIPSON, H. *Nature* **151** (1943) 137.
223. WARREN, B. E. and AVERBACH, B. L. *J. Appl. Phys.* **20** (1949) 1066.
224. WARREN, B. E. and AVERBACH, B. L. *J. Appl. Phys.* **20** (1949) 885.
225. WARREN, B. E. and AVERBACH, B. L. *J. Appl. Phys.* **21** (1950) 595.
226. WARREN, B. E. and AVERBACH, B. L. *J. Appl. Phys.* **23** (1952) 497.
227. WARREN, B. E. and AVERBACH, B. L. In *Modern Research Techniques in Physical Metallurgy*, A.S.M. (1953) 95.
228. BURGERS, W. G. In *Handbuch der Metallphysik*. Akademische Verlagsgesellschaft, Vol. 3, Pt. 2 (1941). esp. 96–107.
229. KOEHLER, J. S. *Phys. Rev.* **60** (1941) 397.
230. KOEHLER, J. S. *Amer. J. Phys.* **10** (1942) 275.
231. SEITZ, F. and READ, T. A. *J. Appl. Phys.* **12** (1941) 100.
231a. SEITZ, F. and READ, T. A. *J. Appl. Phys.* **12** (1941) 170.
232. BRAGG, W. L. *Trans. N.E. Cst Instn Engrs Shipb.* **62** (1945) 25.
232a. BOLLMANN, W. *Phys. Rev.* **103** (1956) 1588.
233. HIRSCH, P. B., HORNE, R. W. and WHELAN, M. J. *Phil. Mag.* **1** (1956) 677.
234. GAY, P. and KELLY, A. *Acta Cryst.* **6** (1953) 165.
235. GAY, P. and KELLY, A. *Acta Cryst.* **6** (1953) 172.
236. GAY, P., HIRSCH, P. B. and KELLY, A. *Acta Cryst.* **7** (1954) 41.
237. KELLY, A. *Acta Cryst.* **7** (1954) 554.
238. HIRSCH, P. B. *Progress in Metal Physics* **6** (1956) 236.
239. SWANN, P. R. In *Electron Microscopy and Strength of Crystals*, ed. by G. Thomas and J. Washburn. Interscience (1963) 131.
240. KEH, A. S. and WEISSMANN, S. In *Electron Microscopy and Strength of Crystals*, ed. by G. Thomas and J. Washburn. Interscience (1963) 231.
241. NABARRO, F. R. N., BASINSKI, Z. S. and HOLT, D. B. *Advances in Physics* **13** (1964) 193.
242. FELTNER, C. E. and LAIRD, C. *Acta Met.* **15** (1967) 1633.
243. STEEDS, J. W. *Proc. Roy. Soc.* **A292** (1966) 343.
244. EMBURY, J. D., KEH, A. S. and FISHER, R. M. *Trans. Met. Soc. AIME* **236** (1966) 1252.
245. LANGFORD, G. and COHEN, M. *Trans. Amer. Soc. Metals* **62** (1969) 623.
246. DINGLEY, D. J. and MCLEAN, D. *Acta Met.* **15** (1967) 885.

247. Holt, D. L. *J. Appl. Phys.* **41** (1970) 3197.
248. Staker, M. R. and Holt, D. L. *Acta Met.* **20** (1972) 569.
249. Edington, J. W. *Phil. Mag.* **19** (1969) 1189.
250. Edington, J. W. *Phil. Mag.* **20** (1969) 531.
251. Peck, J. F. and Thomas, D. A. *Trans. Met. Soc. AIME* **221** (1961) 1240.
252. Johari, O. and Thomas, G. *Acta Met.* **12** (1964) 1153.
253. Nolder, R. L. and Thomas. G. *Acta Met.* **12** (1964) 227.
254. Kressel, H. and Brown, N. *Acta Met.* **14** (1966) 1860.
255. Leslie, W. C., Hornbogen, E., and Dieter, G. E. *J. Iron Steel Inst.* **200** (1962) 622.
256. Smith, C. S. *Trans. Met. Soc. AIME* **212** (1958) 574.
257. Cohen, J. B., Nelson, A. and De Angelis, R. *J. Trans. Met. Soc. AIME* **236** (1966) 133.
258. Jones, R. L. and Conrad, H. *Trans. Met. Soc. AIME* **245** (1969) 779.
259. Edington, J. W. and Smallman, R. E. *Acta Met.* **12** (1964) 1313.
260. Basinski, Z. S. and Basinski, S. J. *Phil. Mag.* **9** (1964) 51.
261. Argon, A. S. and Brydges, W. T. *Phil. Mag.* **18** (1968) 817.
262. Kronmüller, H. *Canad. J. Phys.* **45** (1967) 631.
263. Bollmann, W. *J. Inst. Metals* **87** (1959) 439.
264. Bailey, J. E. In *Proc. European Regional Conf. on Electron Microscopy,* Delft 1960, Vol. 1, 433.
265. Taylor, G. I. *Proc. Roy. Soc.* **A145** (1934) 362.
266. Stroh, A. N. *Proc. Roy. Soc.* **A218** (1953) 391.
267. Mott, N. F. *Phil. Mag.* **63** (1952) 1151.
268. Kuhlmann-Wilsdorf, D. In *Work Hardening,* ed. by J. P. Hirth and J. Weertman, Gordon & Breach (1968) 97.
269. Kuhlmann-Wilsdorf, D. *Met. Trans.* **1** (1970) 3173.
270. Otte, H. M. and Hren, J. J. *Expt. Mech.* **6** (1966) 177.
271. ———
272. Johnson, H. H. *J. Appl. Phys.* **37** (1966) 1763.
273. Seeger, A. and Kronmüller, H. *Phil. Mag.* **7** (1962) 897.
274. Clarebrough, L. M., Hargreaves, M. E., Head, A. K. and Loretto, M. H. *Phil. Mag.* **6** (1961) 819.
275. Moore, J. T. and Kuhlmann-Wilsdorf, D. In *Proc. 2nd Int. Conf. on the Strength of Metals and Alloys,* A.S.M. (1970) 484.
276. Wilkens, M. *Acta Met.* **15** (1967) 1412.
277. Kocks, U. F. and Scattergood, R. O. *Acta Met.* **17** (1969) 1161.
278. Boas, W. In *Defects in Crystalline Solids,* Physical Soc. London (1955) 212.
279. ———
280. Ham, R. K. *Phil. Mag.* **7** (1962) 1177.
281. Fujita, H., Kawasaki, Y., Furubayashi, E., Kajiwara, S. and Taoka, T. *Jap. J. Appl. Phys.* **6** (1967) 214.
282. Stehle, H. and Seeger, A. *Z. Phys.* **146** (1956) 217.
283. Lomer, W. M. *Phil. Mag.* **2** (1957) 1053.
284. Koehler, J. S. In *Impurities and Imperfections.* A.S.M. (1955) 162.
285. Boas, W. In *Dislocations and Mechanical Properties of Crystals,* ed. by J. C. Fisher, W. G. Johnston, R. Thomson and T. Vreeland Jr., Wiley (1957) 333.
286. Seeger, A. In *Dislocations and Mechanical Properties of Crystals,* ed. by J. C. Fisher, W. G. Johnston, R. Thomson and T. Vreeland Jr., Wiley (1957) 347.
287. Basinski, Z. S., Dugdale, J. S. and Howie, A. *Phil. Mag.* **8** (1963) 1989.
288. Cotterill, R. M. J. *Phil. Mag.* **6** (1961) 1351.
289. Hunter, S. C. and Nabarro, F. R. N. *Proc. Roy. Soc.* **A220** (1953) 542.
290. Harrison, W. A. *J. Phys. Chem. Solids* **5** (1958) 44.
291. Seeger, A. and Stehle, H. *Z. Phys.* **146** (1956) 242.
292. Rider, J. G. and Foxon, C. T. B. *Phil. Mag.* **13** (1966) 289.
293. Basinski, Z. S. and Saimoto, S. *Canad. J. Phys.* **45** (1967) 1161.
294. Ziman, J. M. *Principles of the Theory of Solids,* Cambridge (1964).
295. Michell, D. *Acta Met.* **6** (1958) 141.
296. Averbach, B. L., Bever, M. B., Comerford, M. F. and Leach, J. S. Ll. *Acta Met.* **6** (1958) 142.
297. Williamson, G. K. and Smallman, R. E. *Acta Cryst.* **7** (1954) 574.
298. Faulkner, E. A. *Phil. Mag.* **5** (1960) 519.
299. Williamson, G. K. and Smallman, R. E. *Phil. Mag.* **1** (1956) 34.
300. Smallman, R. E. and Westmacott, K. H. *Phil. Mag.* **2** (1957) 669.
301. Mikkola, D. E. and Cohen, J. B. *Acta Met.* **14** (1966) 105.
302. Clarebrough, L. M. and Hargreaves, M. E. *J. Aust. Inst. Metals* **3** (1958) 31.
303. ———
304. Nicholas, J. F. *Phil. Mag.* **46** (1955) 87.

305. ——
306. VAN BUEREN, H. G. *Imperfections in Crystals*, North Holland (1960).
307. DRUYVESTEYN, M. J. and MANINTVELD, J. A. *Nature* **168** (1951) 868.
308. HIRSCH, P. B. In *N.P.L. Symp. No. 15, Teddington 1963*, H.M.S.O. (1963) 240.
309. ZENER, C. *Trans. AIME.* **147** (1942) 361.
310. SEGALL, R. L., PARTRIDGE, P. G. and HIRSCH, P. B. *Phil. Mag.* **6** (1961) 1493.
311. BROOM, T. and MOLINEUX, J. H. *J. Inst. Metals* **83** (1955) 528.
312. BROOM, T. and HAM, R. K. *Proc. Roy. Soc.* **A242** (1957) 166.
313. KUHLMANN-WILSDORF, D. *Trans. Met. Soc. AIME* **224** (1962) 1047.
314. CONRAD, H. and CHRIST, B. In *Recovery and Recrystallization of Metals*, ed. by L. Himmel, Interscience (1963) 124.
315. WOLFENDEN, A. In *Proc. 2nd Int. Conf. on the Strength of Metals and Alloys*, A.S.M. (1970) 489.
316. KUHLMANN-WILSDORF, D. *Scripta Met.* **2** (1968) 643.
317. KRONMÜLLER, H. and WILKENS, M. *Scripta Met.* **3** (1969) 495.
318. KRONMÜLLER, H. and WILKENS, M. *Scripta Met.* **4** (1970) 1.
319. KUHLMANN-WILSDORF, D. and WOLFENDEN, A. *Scripta Met.* **3** (1969) 503.
320. KUHLMANN-WILSDORF, D. and WOLFENDEN, A. *Scripta Met.* **4** (1970) 3.
321. HIRSCH, P. B. *Discussions Faraday Soc.* **38** (1964) 111.
322. LANGE, H. and LÜCKE, K. *Z. Metallk.* **44** (1953) 183 and 514.
323. HSUN HU. *Trans. Met. Soc. AIME* **224** (1962) 75.
324. HSUN HU. *Acta Met.* **10** (1962) 1112.
325. WALTERS, J. L. and KOCH, E. F. *Acta Met.* **11** (1963) 923.
326. SCHRODER, K. *Proc. Phys. Soc. Lond.* **72** (1958) 33.
327. SCHRODER, K. *Proc. Phys. Soc. Lond.* **73** (1959) 674.
328. MERTZ, J. C. and MATHEWSON, C. H. *Trans. AIME* **124** (1937) 59.
329. HOLT, D. L. *Acta Met.* **13** (1965) 39.
330. THORNTON, P. R., MITCHELL, T. E. and HIRSCH, P. B. *Phil. Mag.* **7** (1962) 1349.
331. HOWIE, A. and SWANN, P. R. *Phil. Mag.* **6** (1961) 1215.
332. SMALLMAN, R. E. and GREEN, D. *Acta Met.* **12** (1964) 145.
333. JØSSANG, T. and HIRTH, J. P. *Phil. Mag.* **13** (1966) 657.
334. HÄUSSERMANN, F. and WILKENS, M. *Phys. Stat. Sol.* **18** (1966) 609.
335. LORETTO, M. H. *Phil. Mag.* **12** (1965) 125.
336. STEEDS, J. W. *Phil. Mag.* **16** (1967) 785.
337. FLINN, P. A. In *Strengthening Mechanisms in Solids*, A.S.M. (1962) 17.
338. RUDMAN, P. S. and AVERBACH, B. L. *Acta Met.* **4** (1956) 382
339. HULTGREN, R., ORR, R. L., ANDERSON, P. D. and KELLY, K. K. *Selected Values of Thermodynamic Properties of Metals and Alloys*, Wiley (1963).
340. MARCINKOWSKI, M. J., BROWN, N. and FISHER, R. M. *Acta Met.* **9** (1961) 129.
341. KUHLMANN, D. *Z. Phys.* **124** (1948) 468.
342. KUHLMANN, D., MASING, G. and RAFFELSIEPER, J. *Z. Metallk.* **40** (1949) 241.
343. LI, J. C. M. In *Recrystallization, Grain Growth and Textures*, A.S.M. (1966) 45.
344. JOHNSON, W. A. and MEHL, R. F. *Trans. Amer. Inst. Min. (Metall.) Engrs* **135** (1939) 416.
345. AVRAMI, M. *J. Chem. Phys.* **7** (1939) 1103.
346. AVRAMI, M. *J. Chem. Phys.* **8** (1940) 212.
347. AVRAMI, M. *J. Chem. Phys.* **9** (1941) 177.
348. ANDERSON, W. A. and MEHL, R. F. *Trans. Amer. Inst. Min. (Metall.) Engrs* **161** (1945) 140.
349. COTTRELL, A. H. *Progress in Metal Physics* **4** (1953) 205.
350. CAHN, R. W. *Proc. Phys. Soc. Lond.* **A63** (1950) 323.
351. BAILEY, J. E. *Phil. Mag.* **5** (1960) 833.
352. CRUSSARD, C. *Rev. Métall.* **41** (1944) 133.
353. BECK, P. A. and SPERRY, P. R. *J. Appl. Phys.* **21** (1950) 150.
354. LEIGHLY, H. P., WALKER, H. L. and MARX, J. W. *Trans. Amer. Inst. Min. (Metall.) Engrs* **197** (1953) 809.
355. WALKER, H. L. and BATTACHARYA, D. L. *J. Indian Inst. Sci.* **37B** (1955) 179.
356. MCNEIL, M. B. and LEIGHLY, H. P. J. *J. Appl. Phys.* **37** (1966) 930.
357. OROWAN, E. In *Dislocations in Metals*, ed. by M. Cohen, AIME (1954) 69.
358. CHRISTIAN, J. W. *The Theory of Transformations in Metals and Alloys*, Pergamon (1965).
359. LESLIE, W. C., PLECITY, F. J. and MICHALAK, J. T. *Trans. Met. Soc. AIME* **221** (1961) 691.
360. LESLIE, W. C., PLECITY, F. J. and AUL, F. W. *Trans. Met. Soc. AIME* **221** (1961) 982.
361. ENGLISH, A. T. and BACKOFEN, W. A. *Trans. Met. Soc. AIME* **230** (1964) 396.
362. SPEICH, G. R. and FISHER, R. M. In *Recrystallization, Grain Growth and Textures*, A.S.M. (1966) 563.
363. NICHOLAS, J. F. *Acta Met.* **7** (1959) 544.
364. SEEGER, A., DIEHL, J., MADER, S. and REBSTOCK, H. *Phil. Mag.* **2** (1957) 323.

365. DARKEN. L. S. *Trans. Amer. Soc. Metals* **54** (1961) 599.
366. GIRAUD, R., *Rev. Métall.* **25** (1928) 347.
367. BOCKSTIEGEL, G. and LÜCKE, K. *Z. Metallk.* **42** (1951) 225.
368. EUGÈNE, F. *C.R. Acad. Sci. Paris* **236** (1953) 2071.
369. EUGÈNE, F. *Microtechnic* **8** (1954) 127.
370. GORDON, P. In *Recovery and Recrystallization of Metals*, ed. by L. Himmel, Interscience (1963) 121.
371. FILATOVS, G. J. and SCHWANEKE, A. E. *Rev. Sci. Instrum.* **4** (1971) 447.
372. FILATOVS, G. J. and LEIGHLY, H. P. *Phys. Stat. Sol.* **8** (1971) K43.

(M.S. submitted for publication, January 1972)

AUTHOR INDEX

Ahlers, G. 12, 170
Anderson, C. T. 12, 17, 24, 77, 79, 117, 124, 170
Anderson, P. D. 157, 176
Anderson, W. A. 163, 176
Appleton, A. S. 17, 18, 34, 38, 42, 52, 54–7, 59, 60, 63, 71, 72, 73, 85, 109, 116, 120, 124, 143, 145, 146, 149, 166, 171, 172, 173
Argon, A. S. 129, 153, 175
Åström, H. U. 22, 28, 38, 104, 105, 106, 110, 142, 160–3, 172, 173
Aul, F. W. 165, 176
Averbach, B. L. 30, 58, 107, 114, 125, 140–3, 155, 156, 167, 173–6
Avrami, M. 162, 176
Avsan, O. 22, 172

Backofen, W. A. 165, 176
Bailey, J. E. 22, 32, 33, 36, 48, 73, 84, 103, 118, 119, 124, 128, 130, 132–40, 142, 143, 148, 151, 163, 164, 165, 172, 174, 175, 176
Balluffi, R. W. 14, 90, 143, 144, 170, 173
Basinski, S. J. 129, 153, 175
Basinski, Z. S. 127, 129, 132, 138, 139, 140, 153, 174, 175
Battacharya, D. L. 164, 176
Beardmore, P. 38, 40, 60, 85, 87, 96, 114, 116, 120, 142, 146, 157, 159, 164, 173, 174
Beck, P. A. 164, 176
Bell, F. 22, 36, 40, 47, 77, 93, 102, 113, 145, 161, 162, 172
Berglund, S. 22, 26, 27, 106, 160, 162, 172
Bever, M. B. 5, 9, 16, 18, 20, 22, 26, 27, 28, 30, 32, 34, 38, 40, 42, 45, 47, 49, 51–60, 62, 63, 64, 66, 69, 76, 77, 79, 82–8, 96, 107, 109, 110, 113, 114, 115, 116, 118–22, 124, 125, 140–3, 145, 146, 149, 154–9, 164, 166, 167, 170, 172–5
Boas, W. 125, 126, 138, 139, 174, 175
Bockstiegel, G. 26, 177
Bogachev, I. N. 44, 54, 172
Bogdanov, Ye. S. 8, 170
Bohnenkamp, K. 28, 73, 115, 173
Bollmann, W. 127, 130, 174, 175
Borelius, G. 22, 26, 27, 106, 160, 162, 172
Bragg, W. L. 126, 174
Bridgman, P. W. 10, 14, 15, 170
Brillhart, D. C. 42, 60, 97, 115, 117, 128, 140, 142, 147, 148, 173
Brindley, G. W. 125, 126, 174
Brinkman, J. A. 90, 110, 173
Broom, T. 149, 176
Brown, N. 30, 48, 55, 106, 128, 158, 172, 175, 176
Brydges, W. T. 129, 153, 175
Burgers, W. G. 126, 174

Caglioti, V. 125, 126, 141, 174

Cahn, R. W. 163, 165, 176
Chaikovskii, E. F. 20, 28, 44, 46, 54, 55, 79, 106, 115, 118, 122, 172
Charbonnier, P. 18, 24, 50, 123, 171
Chin, L. L. J. 20, 21, 42, 89, 109, 171
Christ, B. 150, 176
Christian, J. W. 165, 176
Chu, Boa-Teh 15, 170
Chuang, K. C. 40, 96, 110, 122, 143, 145, 173
Clarebrough, L. M. 5, 17–21, 26, 28, 29, 30, 34, 36, 38, 40, 45, 46, 47, 50, 51, 54, 58, 59, 62, 64–7, 73–82, 90–4, 96–104, 106, 107, 108, 110, 113–17, 119–22, 125, 135, 137–40, 142–6, 148–51, 154, 156, 157, 164, 170–5
Cohen, J. B. 32, 42, 45, 52, 60, 85–8, 97, 115–8, 121, 128, 140, 142, 146, 147, 148, 157, 158, 159, 167, 172, 173, 175
Cohen, M. 127, 131, 132, 141, 174
Coleman, B. D. 15, 170
Comerford, M. F. 30, 58, 107, 114, 125, 140–3, 155, 156, 167, 173, 175
Conrad, H. 129, 150, 175, 176
Cotterill, R. M. J. 139, 175
Cottrell, A. H. 12, 13, 133, 163, 165, 170, 176
Crussard, C. 164, 176

Darken, L. S. 166, 176
De Angelis, R. J. 42, 60, 97, 115, 117, 128, 140, 142, 147, 148, 173, 175
De Groot, S. R. 14, 170
Degtiarev, M. M. 26, 44, 46, 62, 82, 118, 124, 172
Denisova, I. K. 44, 54, 172
Diehl, J. 136, 151, 165, 176
Dienes, G. J. 13, 14, 90, 170, 173
Dieter, G. E. 34, 59, 60, 85, 128, 149, 173, 175
Dillamore, I. L. 69, 153, 165, 173
Dillon, O. W. Jr. 15, 18, 36, 38, 40, 65, 171, 173
Dingley, D. J. 127, 129, 174
Druyvesteyn, M. J. 144, 176
Dudarev, E. F. 18, 42, 80–3, 171
Dugdale, J. S. 139, 140, 175

Ebel, H. 20, 21, 40, 172
Edington, J. W. 128, 129, 175
Embury, J. D. 127, 174
English, A. T. 165, 176
Epifanov, G. I. 17, 26, 171
Erdmann, J. C. 17, 18, 36, 40, 42, 56, 63, 82, 83, 145, 146, 149, 166, 171, 172, 173
Eucken, A. 12, 170
Eugène, F. 28, 177

Farren, W. S. 24, 49, 69, 73, 77, 79, 124, 173
Fastov, N. S. 118, 124, 174

Faulkner, E. A. 125, 139, 141, 164, 174, 175
Fedorov, A. A. 26, 44, 46, 49, 54, 77, 79, 82, 118, 124, 172
Feltner, C. E. 127, 129, 150, 174
Filatovs, G. J. 44, 177
Fisher, R. M. 127, 158, 165, 174, 176
Flinn, P. A. 155, 176
Foxon, C. T. B. 139, 175
France, R. W. 22, 24, 172
Friedel, J. 13, 14, 15, 151, 170
Fujita, H. 139, 175
Furubayashi, E. 139, 175

Galy-Aché 18, 24, 50, 123, 171
Gangulee, A. 42, 83, 84, 88, 116, 119, 120, 122, 140, 156–9, 167, 173
Gay, P. 127, 174, 175
Gehlen, P. C. 13, 170
Geiss, W. 125, 174
Giraud, R. 24, 177
Gordon, P. 22, 28, 32, 36, 42, 45, 46, 47, 60–3, 73, 91, 92, 96, 97, 103, 104, 110, 113, 114, 115, 117, 128, 140, 142, 147, 148, 151, 161–5, 172, 173, 177
Grant, N. J. 20, 21, 42, 89, 109, 171
Green, D. 155, 176
Greenfield, P. 28, 30, 52, 53, 58, 59, 82, 83, 85, 109, 143, 156, 157, 172
Grishkevich, A. Ye. 18, 40, 50, 171
Gubkin, S. I. 8, 170
Gurtin, M. E. 15, 170

Haig, F. D. 30, 31, 58, 96, 125, 126, 140–3, 173
Halford, G. R. 18, 20, 21, 42, 65, 66, 67, 99, 114, 150, 171
Ham, R. K. 16, 18, 42, 45, 61, 85, 112, 125, 139, 149, 162, 171, 174, 175, 176
Hargreaves, M. E. 5, 17–21, 26, 28, 29, 30, 34, 36, 38, 40, 45, 46, 47, 50, 51, 54, 58, 59, 62, 64–7, 73–82, 90–4, 96–104, 106, 107, 108, 110, 113–7, 119, 120, 121, 122, 125, 135, 137–40, 142–6, 148–51, 154, 156, 157, 164, 170–5
Harrison, W. A. 139, 140, 175
Häussermann, F. 155, 176
Haworth, F. E. 125, 126, 141, 174
Head, A. K. 28, 30, 64, 65, 66, 98, 99, 113, 114, 135, 137, 149, 173, 175
Henderson, J. W. 20, 21, 30, 55, 93, 94, 95, 107, 143, 144, 172
Hertsriken, S. D. 20, 34, 44, 46, 47, 48, 76, 77, 79, 80, 81, 118, 124, 172
Hill, R. 15, 171
Hirn, G. A. 123, 174
Hirsch, P. B. 22, 32, 33, 36, 48, 73, 84, 103, 118, 119, 124, 127, 128, 132–40, 142, 143, 148, 149, 151, 153, 154, 155, 163, 164, 165, 172, 174, 175, 176
Hirth, J. P. 155, 176
Holt, D. B. 127, 132, 138, 174
Holt, D. L. 127, 154, 155, 174, 175, 176
Holtzman, A. H. 40, 60, 85, 87, 114, 142, 146, 157, 159, 173
Hornbogen, E. 128, 175

Horne, R. W. 127, 174
Hort, H. 18, 23, 24, 48, 115, 118, 123, 124, 171
Howie, A. 139, 140, 155, 175, 176
Howlett, B. W. 20, 22, 172
Hren, J. J. 133, 175
Hsun Hu 153, 176
Hultgren, R. 157, 176
Hunter, S. C. 139, 140, 175

Iyer, A. S. 32, 36, 60, 62, 63, 96, 97, 113, 114, 148, 173

Jahoda, J. A. 17, 18, 36, 40, 42, 56, 63, 82, 83, 145, 146, 149, 166, 171, 172, 173
Johari, O. 128, 175
Johnson, H. H. 134, 175
Johnson, W. A. 162, 176
Jones, R. L. 129, 175
Jøssang, T. 155, 176

Kajiwara, S. 139, 175
Kalinina, E. N. 20, 32, 171
Kanzaki, H. 20, 26, 44, 48, 49, 51, 55, 69, 77, 91, 99, 103, 171
Kawasaki, Y. 139, 175
Keh, A. S. 127, 128, 129, 149, 150, 151, 174
Kelly, A. 127, 174, 175
Kelly, K. K. 157, 176
Kestin, J. 14, 15, 170
Khotkevich, V. I. 12, 20, 28, 34, 44, 46, 48, 54, 55, 79, 88, 89, 106, 115, 118, 119, 122, 170, 172, 173, 174
Koch, E. F. 153, 176
Kocks, U. F. 136, 175
Koehler, J. S. 20, 21, 30, 55, 90, 93, 94, 95, 107, 126, 131, 133, 134, 139, 143, 144, 172–5
Koref, F. 22, 172
Korenchenko, Ye. S. 18, 40, 50, 171
Kovacs, I. 20, 21, 42, 93, 103, 105, 171
Koyama, K. 19, 20, 38, 45, 46, 66, 73, 74, 75, 99, 100, 154, 171
Kratochvil, J. 15, 171
Kressel, H. 128, 175
Krisement, O. 22, 36, 47, 77, 93, 113, 145, 161, 162, 172
Krivobok, V. N. 20, 24, 69, 106, 151, 172
Kronmüller, H. 130, 135, 136, 137, 151, 152, 153, 175, 176
Kuhlmann, D. 160, 176
Kuhlmann-Wilsdorf, D. 131, 132, 136, 137, 150, 151, 152, 175, 176
Kunin, N. F. 18, 24, 40, 44, 46, 50, 118, 124, 171, 172
Kunin, V. N. 18, 32, 40, 44, 48, 50, 77, 79, 115, 118, 119, 122, 171

Laird, C. 127, 129, 150, 174
Lange, H. 153, 176
Langford, G. 127, 131, 132, 141, 174
Larikov, L. N. 20, 34, 44, 46, 47, 48, 76, 77, 79, 80, 81, 118, 124, 172

Leach, J. S. Ll. 16, 22, 28, 30, 52, 58, 107, 114, 125, 140–3, 155, 156, 167, 171, 172, 173, 175
Leighly, H. P. 44, 164, 176, 177
Leslie, W. C. 128, 165, 175, 176
Li, J. C. M. 13, 142, 162, 165, 170, 176
Lipson, H. 125, 174
Loewen, E. G. 28, 52, 143, 172
Lomer, W. M. 139, 175
Loretto, M. H. 5, 20, 30, 34, 36, 38, 40, 45, 46, 47, 54, 62, 66, 67, 73–81, 90, 93, 94, 96–104, 106, 107, 108, 110, 113–7, 119, 122, 125, 135, 137–40, 142–6, 148, 149, 150, 154–7, 164, 170, 172–6
Lovegrove, E. 32, 73, 140, 141, 142, 173
Lücke, K. 26, 28, 73, 115, 153, 173, 176, 177
Lugscheider, W. 20, 21, 40, 42, 55, 96, 105, 145, 172

McAdam, D. J. Jr. 124, 174
McLean, D. 127, 129, 174
McNeil, M. B. 164, 176
Mader, S. 136, 151, 165, 176
Maier, C. G. 12, 17, 24, 77, 79, 117, 124, 170, 174
Manintveld, J. A. 144, 176
Marcinkowski, M. J. 158, 176
Marshall, E. R. 28, 45, 47, 49, 172
Martin, D. L. 12, 170
Marx, J. W. 164, 176
Masima, M. 24, 49, 69, 172
Masing, G. 28, 73, 115, 160, 173, 176
Mathewson, C. H. 154, 176
Mazur, P. 14, 170
Meechan, C. J. 90, 110, 173
Mehl, R. F. 162, 163, 176
Meixner, J. 15, 170
Mertz, J. C. 154, 176
Michalak, J. T. 165, 176
Michell, D. 19, 21, 26, 30, 31, 32, 45, 46, 50, 58, 73, 96, 125, 126, 140–3, 171, 173, 175
Mikkola, D. E. 142, 146, 147, 157, 158, 159, 167, 175
Milevskaya, V. G. 18, 30, 44, 47, 77, 82, 119, 171
Mima, G. 20, 32, 93, 96, 103, 104, 142, 171
Misra, S. 20, 172
Mitchell, T. E. 155, 176
Molineux, J. H. 149, 176
Moore, J. T. 136, 137, 152, 175
Mott, N. F. 131, 134, 175

Nabarro, F. R. N. 127, 132, 138, 139, 140, 174, 175
Nakada, Y. 18, 40, 70–3, 118, 119, 124, 153, 166, 171
Nelson, A. 128, 175
Nicholas, J. F. 66, 110, 143, 161, 163–6, 173, 175, 176
Nolder, R. L. 128, 175

Okamoto, M. 20, 21, 40, 69, 73, 75, 84, 99, 102, 106, 109, 114, 116, 151, 153, 157, 172
Orowan, E. 164, 176
Orr, R. L. 157, 176
Otte, H. M. 133, 175

Panin, V. E. 18, 30, 42, 44, 47, 77, 80–3, 119, 171
Partridge, P. G. 149, 176
Pascoe, K. J. 125, 174
Peck, J. F. 128, 175
Pervakov, V. A. 12, 34, 48, 115, 118, 119, 122, 170, 174
Plecity, F. J. 165, 176
Polakowski, N. H. 62, 173
Popov, M. M. 20, 32, 171
Preban, A. G. 42, 60, 97, 115, 117, 128, 140, 142, 147, 148, 173
Prigogine, I. 14, 170

Quinney, H. 17, 20, 24, 26, 44, 46–50, 77, 80, 81, 103, 106, 118, 124, 126, 171

Raffelsieper, J. 160, 176
Read, T. A. 126, 131, 133, 174
Rebinder, P. A. 17, 26, 171
Rebstock, H. 136, 151, 165, 176
Regnault, V. 12, 170
Reik, H. G. 15, 170
Rice, J. R. 15, 170, 171
Richards, C. E. 69, 173
Rider, J. G. 139, 175
Ridley, P. 125, 126, 174
Riggs, F. B. Jr. 28, 55, 172
Robinson, P. M. 40, 55, 87, 88, 116, 120, 158, 172, 173
Rocca, R. 9, 170
Roessler, B. 34, 52, 85, 86, 116, 121, 143, 157, 172
Rosenhain, W. 17, 24, 77, 79, 123, 124, 171, 174
Rubin, D. 15, 170
Rudman, P. S. 156, 176
Russell, T. F. 123, 174

Sachs, G. 24, 49, 69, 125, 126, 141, 172, 174
Saimoto, S. 140, 175
Sato, S. 19, 20, 23, 24, 46, 48, 49, 50, 77, 79–83, 91, 103, 105, 106, 107, 157, 171
Scattergood, R. O. 38, 60, 96, 114, 136, 164, 173, 175
Schottky, W. F. 32, 77, 79, 84, 173
Schroder, K. 154, 176
Schwaneke, A. E. 44, 177
Seeger, A. 13, 133, 135, 136, 137, 139, 140, 151, 152, 153, 155, 165, 170, 175, 176
Segall, R. L. 40, 62, 73, 74, 75, 97, 98, 100, 117, 138, 142, 148, 149, 154, 173, 176
Seitz, F. 126, 131, 133, 174
Senilov, G. W. 24, 44, 46, 118, 124, 172
Shepelev, A. G. 34, 48, 115, 118, 119, 122, 174
Shermergor, T. D. 18, 30, 171
Simmons, R. O. 14, 90, 143, 144, 170, 173
Sinnatt, –. 24, 123
Sirenko, G. A. 44, 79, 88, 89, 173
Sizmann, R. 20, 32, 172
Sjöberg, S. 22, 26, 27, 106, 160, 162, 172
Skuratov, S. M. 20, 32, 171
Slyusar, B. P. 20, 34, 44, 46, 47, 48, 76, 77, 79, 80, 81, 118, 124, 172

Smallman, R. E.   129, 141, 142, 155, 164, 175, 176
Smith, C. J.   22, 24, 172
Smith, C. J. E.   69, 153, 165, 173
Smith, C. S.   125, 128, 174, 175
Smith, J. H.   42, 45, 52, 77, 79, 109, 115, 119, 120, 122, 172
Speich, G. R.   165, 176
Sperry, P. R.   164, 176
Staker, M. R.   127, 175
Steeds, J. W.   127, 129, 155, 174, 176
Stehle, H.   139, 140, 175
Stibitz, G. R.   125, 141, 174
Stickley, E. E.   125, 174
Stokes, A. R.   125, 174
Stott, V. H.   17, 24, 77, 79, 124, 171
Stroh, A. N.   131, 134, 135, 175
Studenok, Iu. A.   26, 44, 46, 58, 59, 61, 82, 118, 124, 172
Sukhovarov, E.   18, 42, 80-3, 171
Suzuki, K.   20, 21, 40, 69, 73, 75, 84, 99, 102, 106, 109, 114, 116, 151, 153, 157, 172
Suzuki, T.   12, 20, 26, 44, 46, 91, 113, 118, 124, 170
Swann, P. R.   127, 155, 174, 176

Tammann, G.   10, 170
Taoka, T.   20, 21, 40, 69, 73, 75, 84, 99, 102, 106, 109, 114, 116, 139, 151, 153, 157, 172, 175
Taylor, G. I.   17, 20, 24, 26, 44, 46-50, 69, 73, 77, 79, 80, 81, 103, 106, 118, 124, 125, 126, 131, 171, 173, 174, 175
Thomas, D. A.   128, 175
Thomas, G.   128, 175
Thomson, W. (Lord Kelvin)   9, 170
Thornton, P. R.   155, 176
Ticknor, L. B.   22, 26, 27, 28, 45, 47, 49, 58, 172
Tikhomina, E. N.   20, 32, 171
Titchener, A. L.   5, 16, 18, 20, 30, 32, 45, 47, 49, 51, 52, 53, 56-9, 62, 64, 66, 69, 85, 109, 113, 143, 146, 149, 154, 170, 172, 173
Tizhnova, N. V.   26, 44, 46, 81, 82, 118, 124, 172
Tokizawa, M.   20, 32, 93, 96, 103, 104, 142, 171

Van Bueren, H. G.   144, 176
Van den Beukel, A.   20, 21, 34-8, 55, 84, 94, 95, 103, 105, 107, 109, 143, 144, 145, 154, 172

Vandermeer, R. A.   36, 73, 104, 161, 162, 163, 165, 173
Van Liempt, J. A. M.   22, 125, 172, 174
Vineyard, G. H.   13, 14, 90, 170, 173

Waldman, J.   42, 83, 116, 119, 122, 124, 173
Walker, H. L.   164, 176
Walters, J. L.   153, 176
Wang, T. P.   30, 48, 55, 106, 172
Warren, B. E.   125, 140, 141, 174
Warrentrup, H.   10, 170
Watson, T. W.   69, 153, 165, 173
Webeler, R.   20, 28, 63, 64, 171
Weissmann, S.   127, 128, 129, 149, 150, 151, 174
Welber, B.   20, 26, 28, 63, 64, 171
Wenzl, H.   20, 32, 38, 74, 76, 100, 154, 163, 172
West, G. W.   17, 18, 19, 21, 26, 28, 29, 30, 34, 45, 46, 47, 50, 51, 58, 59, 64, 65, 66, 73-9, 82, 91, 92, 93, 96, 98-103, 113-7, 119-22, 138, 139, 142, 143, 145, 149, 151, 154, 171, 173, 174
Westmacott, K. H.   142, 175
Whelan, M. J.   127, 174
White, A. J.   34, 66, 74, 99, 100, 173
White, J. L.   19, 20, 36, 38, 45, 46, 66, 73, 74, 75, 99, 100, 151, 154, 171, 172
Wildhack, H.   20, 21, 42, 55, 96, 105, 145, 172
Wilkens, M.   136, 151, 152, 155, 175, 176
Williams, R. O.   5, 16, 17, 18, 34, 36, 38, 40, 44-52, 54, 58, 61, 66-81, 84, 110, 111, 112, 118, 119, 124, 151, 153, 154, 162, 164, 170, 171, 173
Williamson, G. K.   141, 142, 164, 175
Wolfenden, A.   17, 18, 42, 44, 46, 56, 68, 69, 71, 72, 73, 85, 87, 118, 119, 124, 151, 152, 171, 172, 173, 176
Wolff, H.   22, 172
Wood, W. A.   125, 174

Yoshikawa, A.   20, 21, 40, 69, 73, 75, 84, 99, 102, 106, 109, 114, 116, 151, 153, 157, 172

Zashkvara, V. V.   20, 28, 44, 46, 54, 55, 79, 106, 115, 118, 122, 172
Zener, C.   148, 176
Ziman, J. M.   140, 175

SUBJECT INDEX

Activation energy 110, 143, 144
  of recovery 106, 160–1
  of recrystallization 163–4
  of self-diffusion 14–15
Alloys
  age hardened 169
  dispersion hardened 89, 169
  long-range ordered 85–8, 157–9, 169
    structure of cold-worked 130
  multi-phase 169
  release of stored energy from 106–9
  solid solution 155–7, 168, 169
  stored energy in 80–9
    effect of impurities 76
    effect of long-range order 85–8
  stored energy/expended energy ratio 82–3
  two-phase 88–9
  see also under specific alloys
Aluminum
  dislocation density in 138–9
  release of stored energy from 103–5
    after low-temperature deformation 95–6, 105, 145
    during recovery 95–6, 103–5, 110, 115
    during recrystallization 96, 103–5, 110, 115
    immediately after deformation 110–1, 162
    kinetics of recovery 160–1
    kinetics of recrystallization 163
  stored energy in 75, 77–80
    against expended energy 49
    against extent of deformation 47
    against temperature of deformation 55, 56
    effect of cyclic loading 65
    effect of purity 75
    in single crystals 69, 70–3, 153, 166
    in solid solutions 83
    related to
      density 117
      flow stress 119
      hardness 113
      melting temperature 78–9
      nuclear magnetic resonance 125
      resistivity 115, 122
      stacking fault energy 154–5
  stored energy/expended energy ratio 51, 65, 69
Aluminum bronze
  see Copper-aluminum alloys
Aluminum-copper alloys
  kinetics of recovery 161
  kinetics of recrystallization 163–4
Aluminum-silver alloys
  dislocations in 139
Amorphous metal theory 123
Annealing 12, 18, 19, 21, 148
  anisothermal, heating rate in 110
  immediately after deformation 110–2

  isochronal, temperature in 110
  isothermal, temperature in 110
  of point defects 144–6
Antiphase boundaries 130, 147, 157–9
  see also Destruction of order
Arsenical copper
  release of stored energy from 100, 107, 116, 154
  stored energy in 74
    related to hardness 113
    related to resistivity 116

Bauschinger effect 62, 113, 129, 148–9
Bismuth-thallium alloy
  stored energy in 88, 158
    related to resistivity 116, 120–1, 158
Brass
  see Copper-zinc alloys
Bronze
  see Copper-tin alloys, Copper-aluminum alloys

Cadmium
  release of stored energy from 106
  stored energy in
    against identity of metal 77, 79
    against temperature of deformation 54, 56
    related to resistivity 118, 122
Cadmium-lead alloys
  see Lead-cadmium alloys
Calorimetry 16–22
Cell structure 127–33, 135, 140–2, 146, 150, 154
Cell walls
  see Cell structure
Clustering 168
Cold work
  definition 6
  stored energy of
    see Stored energy
Cold-worked metals
  relation between variables and structure of 127–30
Compression, deformation by 8, 9, 10, 23, 73, 99, 124, 128, 146
  in aluminum 70–1, 96, 103–4
  in cadmium 79
  in copper 17–8, 51, 55, 59, 61, 62, 66–9, 74, 95, 96–7, 100, 113, 117, 119, 149
  in copper-gold alloys 84, 107
  in copper-nickel alloys 82
  in copper-silver alloys 84, 107
  in copper-zinc alloys 55, 107
  in gold 103
  in lead 123
  in lead-cadmium alloys 88–9
  in molybdenum 106

SUBJECT INDEX

in nickel 117, 141
in silver 71, 103
in single crystals 69, 70–1
Copper
 Bauschinger effect in 62, 113, 148
 dislocation density in 138–40, 148, 151
 release of stored energy from 91–102
  after cyclic deformation 98–9
  after irradiation 144
  after low-temperature deformation 93–5, 143–4, 146
  during recovery 91, 93–5, 97–100, 114–5, 117, 142, 144, 148–9
  during recrystallization 91, 93, 96–8, 100, 103, 114–5, 144, 149, 157
  effect of deformation history 97–8
  effect of purity 100, 107
  immediately after deformation 85, 110, 112
  kinetics 160, 163–4
 stored energy in
  against
   deformation process 59–61, 147–8
   expended energy 23, 46, 80
   extent of deformation 23, 50, 72, 74, 124
   identity of metal 76–80
   rate of deformation 58, 59, 61
   temperature of deformation 51, 54–6
  effect of
   alumina 89
   cyclic deformation 63–6, 149
   deformation history 148
   grain size 66–8, 150
   incremental deformation 61
   purity 73–4, 76
  in single crystals 69–72, 151
  related to
   density 117, 121, 140
   flow stress 119
   hardness 113–4
   nuclear magnetic resonance 125
   resistivity 115–6, 118–20, 122, 140
   stacking fault energy 154–5
  spread of values 23, 73, 77, 168
 stored energy/expended energy ratio 50–2, 56, 72, 81, 151–2, 153
  effect of purity 74
  in cyclic deformation 65–6
 thermoelastic effect in 10
 work softening in 65, 113–4
Copper-alumina
 release of stored energy from 109
 stored energy in 89
Copper-aluminum alloys
 release of stored energy from 106
  immediately after deformation 111–2, 162
 stored energy in 83–4
Copper-beryllium alloy
 release of stored energy from 107
Copper-gold alloys
 release of stored energy from 107, 144–5
 stored energy in 84–7, 157–9
  against deformation process 60, 146–7

against extent of deformation 45, 85–7
against flow stress 118
against temperature of deformation 52, 55
related to resistivity 116, 121, 157
stored energy/expended energy ratio 86
Copper-nickel alloys
 dislocation structure in 146
 release of stored energy from 106
 stored energy in
  against composition 81–3
  against temperature of deformation 56
  effect of impurities 76
 stored energy/expended energy ratio 56, 82–3
 work softening in 63, 149
Copper-silver alloys
 release of energy from 107, 144–5, 154
 stored energy in
  against composition 84
  against temperature of deformation 55
  effect of purity 76
Copper-silver-nickel alloy
 release of stored energy from 157
Copper-tin alloys
 release of stored energy from 106
  immediately after deformation 112, 162
 stored energy/expended energy ratio 85
Copper-zinc alloys
 dislocation density in 139
 release of stored energy from 106–7
  during recovery 107, 114, 116, 118, 157
  during recrystallization 107, 114, 116, 118, 157
  effect of heating rate 110
  immediately after deformation 111–2, 162
 stored energy in 20
  against composition 80–3
  against temperature of deformation 55
  effect of purity 76
  in single crystals 69
  related to
   density 117–8
   destruction of short-range order 156–7
   hardness 114
   resistivity 115, 122
 stored energy/expended energy ratio 81
Copper-zinc-nickel alloy
 release of stored energy from 106
Cyclic deformation 8, 18, 23, 113–4, 129, 149–50
 dislocation structure after 129
 hardness changes in 113–4
 release of energy after 98–9, 149–50
 stored energy after 63–6

Deformation
 cyclic see Cyclic deformation
 dissipation of energy during 165–6
 effect of temperature of specimen 9–10, 16–7, 65
 elastic see Elastic deformation
 extent of 23–51, 91–3, 127–8, 130–1
  see also under specific materials
 history effect of see path of
 path of 8, 58, 61–3, 97–8, 129, 148–9, 169
 plastic see Plastic deformation

SUBJECT INDEX

process of 59–61, 96–7, 128, 146–8
  see also under specific deformation processes
rate of 56–9, 96, 128, 146
release of stored energy immediately after 110–2, 162
  see also under specific materials
temperature of 51–6, 93–6, 128, 143–6
thermal effects of 9–10
thermodynamic aspects of 6–16
thermodynamic function change with 10–2
work of 7–9, 66
Deformation process variables
  effect on amount of stored energy 23–66
  effect on release of stored energy 91–9
Density 97, 117–8, 124, 142, 145, 148, 156
  dislocation density from 138
  measurements 117–8
  relation to resistivity 140
  relation to stored energy 118–22, 138, 140, 142, 148
Destruction of order
  contribution to stored energy 155–9, 168, 169
Differential power analysis 19, 20, 21
Differential thermal analysis 19, 20
Dislocation densities 15, 126–30, 132–3, 146, 147, 148, 161, 168, 169
  and flow stress 133, 136
  and grain size 150–1
  and recovery 142–3, 145–6, 148, 149, 153–4
  and stored energy 133–8, 150–1
  and strain energy 133–5
  estimation of 138–40
  from X-ray line broadening 142
  in alloys 157
    with long-range order 157
Dislocation structure of cold-worked metals 127–30
  after cyclic deformation 149–50
  and density change 148–9
  and recovery 142–3
    immediately after deformation 162
  and stored energy 132–40
  in single crystals 151–3
Dislocation theory 127
Dislocations 12–3
  and dissipation of energy of deformation 165–6
  and impurity atoms 153
  contribution to stored energy 159, 167, 169
  early models 126
  free energy 13
  interaction energy 13, 135–7, 143, 152–3
  resistivity 139–40
  strain energy 12–3, 131, 133, 148–9
Dispersion hardened alloys 89, 109, 169
Dissipation of energy during deformation 165–6
Drilling, deformation by 18, 52, 58, 59, 83, 109, 156

Easy glide 10, 153
Elastic deformation 6, 10
  thermal effects of 9
Electrical conductivity 18

Electron microscopy 125–8, 130, 132–3, 139–42, 146–9, 151, 153, 164
Energy of imperfections 12–4
Energy storage, incremental rate of
  see Rate of energy storage
Energy storage, instantaneous rate of
  see Rate of energy storage
Enthalpy change 11
Entropy change 11–2
Entropy of imperfections 12–3
Expended energy 9, 17–8, 19
  see also Stored energy/expended energy ratio
Explosive loading, deformation by
  see Shock loading, deformation by
Extension, deformation by
  see Tension, deformation by

Fatigue *see* Cyclic deformation
Filing, deformation by 58, 107–9, 114, 140, 141, 156
First Law of Thermodynamics 6, 14, 16
Flow stress
  measurements of 115
  relation to
    antiphase boundaries 147
    dislocation density 131–3, 136
    grain size 150
    stored energy 109, 115, 118–22, 151
Free energy 10–1
Friction work 8, 16, 61

Gibbs free energy 10–1
Gold
  dislocation densities in 139–40
  release of stored energy from 103, 110
    after low-temperature deformation 95–6, 103, 144–5
    during recrystallization 103, 115, 117
  stored energy in
    against extent of deformation 45
    against identity of metal 77–9
    against temperature of deformation 55
    relation to
      density 117
      flow stress, 119
      hardness 113
      resistivity 115, 119–20, 122
      stacking fault energy 154–5
Gold-silver alloys
  release of stored energy from 96–7
    after low-temperature deformation 107–9, 146
    during recovery 96, 107–9, 114, 116
    during recrystallization 96, 107–9, 114, 116
    kinetics of recrystallization 164
  stored energy in
    against
      deformation process 59–60, 146
      expended energy 47, 49, 51
      extent of deformation 45
      rate of deformation 56–8
      temperature of deformation 51–3, 55, 143

effect of composition 82–3, 85
effect of grain size 66
effect of purity 76
relation to
  destruction of short-range order 155–7
  hardness 113–4
  lattice strain 141
  resistivity 116, 120, 122
  subgrain size 140–1
stored energy/expended energy ratio 50
work softening 62–4, 113
Grain boundary migration 163–4
Grain size 99, 113, 129, 150, 163
  effect on stored energy 66–9
Grain structure 66–9, 99, 129, 150
Grinding, deformation by 58, 96, 140–1

Hardness
  relation to grain size 150
  relation to rate of deformation 146
  relation to stored energy 96, 109, 113–4, 147, 149, 157
Heat capacity 12
Heat effect 6–7, 11, 16–7, 18, 22, 112
Hypothetical temperature rise 125

Impact, deformation by 16, 18, 58, 66, 75, 78, 84–5, 96, 103–4, 110, 112
  see also Shock loading
Imperfections 126–7
  contributions to stored energy 166–8
  energy of 12–4
  entropy of 13–4
Impurities, effect of 47, 73–6, 100–2, 130, 150, 153–4, 163–4, 168–9
Incremental deformation 59, 61, 63, 97, 113–4, 117, 124, 148–9
Intermetallic compounds
  stored energy in 85–8, 168
  structure of cold-worked 130
  see also under specific alloy systems
Internal energy 10, 15
Interstitials 14, 131, 144, 145
  see also Point defects
Intrinsic strength, theory of 124
Iron 123, 131
  dislocation structure of 129, 132, 149
  release of stored energy from 105–6
    during recrystallization 157, 165
    effect of purity 102
    immediately after deformation 110–1, 162
  stored energy in
    against
      expended energy 48
      extent of deformation 23, 47
      identity of metal 77–80
      temperature of deformation 55
    effect of purity 75–6
      related to flow stress 119
    stored energy expended energy ratio 51

Iron-silicon alloys
  release of stored energy from 109
    during recovery 157
    during recrystallization 114, 153, 157
    in single crystals 99, 153
  stored energy in 84–5
    in single crystals 69–70, 153
    relation to
      destruction of short-range order 157
      hardness 114
      resistivity 116
Irradiation 144
Irreversible thermodynamics and plastic deformation 14–6

Kinetics *see* Release of the stored energy

Lattice imperfections, stored energy theories based on 126–7
Lattice strain
  and stored energy 141–3
  stored energy theories based on 125–6
Lead 123
  release of stored energy from 106
    immediately after deformation 110–1, 162
  stored energy in
    against identity of metal 77–9
    against temperature of deformation 54, 56
    relation with resistivity 118, 122
  stored energy/expended energy ratio 51
Lead-cadmium alloys
  stored energy in 88–9
    effect of composition 89
  stored energy/expended energy ratio 89
Lead-indium alloy
  work softening in 63
Long-range ordered alloys 85–8, 157–9
  effect of cold work on
    resistivity 116
    structure 130
    stored energy 85–8

Mechanical equivalent of heat
  relation to stored energy 123
Metal
  identity of
    effect on release of stored energy 103–6
    effect on stored energy 76–80, 154–5
    effect on structure after cold work 130
    purity of 73–6, 100–2, 130, 153–4
  see also Impurities, effect of
Molybdenum
  kinetics of recovery 161
  release of stored energy from 106

Nickel
  dislocation densities in 138–9
  release of stored energy from 90
    after cyclic deformation 99
    after low-temperature deformation 95–6, 145

during recovery 91–3, 95, 101–2, 115, 117, 121, 142
during recrystallization 91–3, 102, 115, 117
effect of deformation process 96
effect of extent of deformation 91–3
effect of purity 101–2
immediately after deformation 110–1, 162
kinetics of recovery 161
stored energy in
against
deformation process 59
expended energy 47, 80
extent of deformation 47
identity of metal 77–80, 82
rate of deformation 58
temperature of deformation 55, 143
effect of cyclic deformation 65, 149–50
effect of purity 74–6
relation to
density 117, 121
flow stress 119
hardness 113
lattice strain 141–2
resistivity 115, 122
stacking fault energy 154–5
subgrain size 140–1
stored energy/expended energy ratio 50–1
thermoelastic effect in 10
Niobium 128
Normalized recrystallization energies 155
Nuclear magnetic resonance, relation to stored energy 125

"Omega phase" 124
Optical microscopy 146
Orthogonal cutting, deformation by 47, 128

Path of deformation *see* Deformation, path of
Permanent "set" 6
Phase change interpretation of stored energy 123–4
Phosphor bronze *see* Copper-tin alloys
Plastic deformation 5–6, 10, 12
and irreversible thermodynamics 14–6
ideal 10
microscopic and macroscopic aspects of 16
thermal effects of 9–10
Plastic strain rate 15
Point defects 12, 130–1, 151, 159
and cyclic deformation 149–50
and dissipation of energy of deformation 165–6
and recovery 148, 156, 161
immediately after deformation 162
annealing of 144–6
contribution to stored energy 130, 167, 169
energies of formation, migration 14–5
entropy of 13–4
energy of 13–4
Preferred orientation 169
Prerecrystallization energy
ratio of total stored energy 90–2, 96, 98, 103, 105

Prerecrystallization processes
see Recovery
Purity of metal 73–6, 130, 153–4
effect on release of stored energy 100–2
see also Impurity effects

Rate of deformation 56–9, 96, 128, 146
Rate of energy storage
incremental 50, 51, 56, 69, 72, 73, 81, 118, 124, 131, 153
instantaneous 50, 118, 124, 130, 137–8, 151–3
Recovery
and density 117–8, 121, 142
and hardness 113–4, 142
and purity 153–4
and resistivity 115–6, 142
and stored energy 90, 123, 142–6, 148, 150, 160–1, 164–5, 167
in alloys 155, 157
kinetics 160–1, 162, 165
overlap with other restoration processes 157
relation to cell structure 128
see also under specific materials
Recrystallization
and density 117, 118, 121
and hardness 113–4
and grain boundary migration 164
and grain size 150
and purity 150, 153–4
and resistivity 115–6
and stacking fault energy 155
and stored energy 90, 123, 130, 143, 162–4, 165, 167
at twin intersections 128
kinetics 162–5
overlap with other restoration processes 157
see also under specific materials
Redundant work 8, 61, 146
Release of the stored energy 89–112
after cyclic deformation 98–9
during prerecrystallization processes 90, 142–6
during recrystallization 90
effect of
deformation process 96–7
extent of deformation 91–3
grain size 99
heating rate 110
identity of the metal 103–6
impurities 100–2, 153–4
rate of deformation 96, 146
temperature of annealing 110
temperature of deformation 93–6, 103, 105, 107, 109, 143–6
from alloys 106–9
from single crystals 99
immediately after deformation 90, 110–2, 153–4, 155, 162
in beginning of isothermal annealing 21–2
kinetics of 20, 21, 90, 123, 153, 160–5
measurement of 89–90
see also Recovery, Recrystallization *and under specific materials*

SUBJECT INDEX

Resistivity
  estimation of dislocation densities from 138–40
  measurements of 115–6
  relation to density 140
  relation to stored energy 97, 109, 115–6, 118–22, 138, 140, 142, 145, 157–8
Restoration processes 5, 164
  see also Recovery, Recrystallization, Release of the stored energy
Rolling, deformation by 8, 23, 61, 74
  in aluminum 125
  in copper 114, 125
  in copper-gold alloys 85, 157
  in gold-silver alloys 47
  in iron 75
  in iron-silicon alloys 84–5, 99, 109, 153
  in single crystals 69–70, 99

Saturation level of stored energy 50, 51, 169
Second Law of Thermodynamics 14
Shear strain, stored energy as function of 69–73
Shock loading, deformation by 6, 23, 59–61, 96–7, 128
  in copper 114–5, 117, 140, 142, 147–8
  in copper-gold alloys 87, 114, 142, 146–7, 157, 159
  in gold-silver alloys 114, 164
Short-range order *see* Alloys, solid solution, Antiphase boundaries, Destruction of order
Silver
  dislocation density in 136, 138–9, 140
  dislocation structure in 134
  recovery in 142
  release of stored energy from
    after low-temperature deformation 95, 103, 144
    during recovery 95, 103
    during recrystallizaion 103, 115, 117
    immediately after deformation 110–1, 162
    kinetics of 163–4
  stored energy in
    against
      expended energy 48
      extent of deformation 47
      identity of metal 77–80
      temperature of deformation 55
    in single crystals 71–3, 152, 166
    relation to
      density 117
      flow stress, 119, 132–3
      hardness 113
      resistivity 115, 118–9, 122
      stacking fault energy 154–5
    stored energy/expended energy ratio 51, 71, 151–2, 166
Silver-cadmium alloys
  stored energy in 83
    relation to resistivity 116, 119, 122
Silver-magnesium alloys
  dislocation density in 140

stored energy in
  against extent of deformation 55, 83–4, 87–8
  against temperature of deformation 55, 87–8
  effect of composition 83–4
  effect of degree of order 88, 156–8
  relation to resistivity 116, 119, 122, 140
Single crystals
  aluminum 56, 69–73, 153
  copper 51, 56, 69, 71–2, 151–3
  copper-zinc alloys 49, 69
  dislocation structure of 129–30, 151–3
  iron-silicon alloys 69–70, 116
  point defects in 151
  release of energy from 99
  silver 71, 72–3, 151–2
  stored energy in 69–73, 169
  stored energy/expended energy ratio 56, 69–73, 151–3
Solid solutions *see* Alloys
Stacking faults 12, 130, 155, 169
  contribution to stored energy 167–8
  resistivity of 139
Steel
  stored energy in 23, 77, 153
  thermoelastic effect in 10
Stored energy (of cold work) 5–6
  amount of 23–89
    against
      deformation history 61–6, 148–9
      deformation process 59–61, 146–8
      expended energy 23–49, 79, 130–1,
      extent of deformation 23–49
      grain size 66–9, 150–1
      identity of metal 76–80
      melting temperature 78–80, 154–5
      purity 73–6
      rate of deformation 56–9, 146
      strain 23, 49, 104–5, 132
      temperature of deformation 20, 22, 51–6, 143–6
    effect of long-range order 85–9, 157–9, 168
    effect of pre-existing substructure 69
    effect of preferred orientation 69
  and cell structure 140–1
  and dislocation density 132–40, 142
  and dislocations 126–7, 132–8, 141, 142, 167
  and lattice strain 125, 131–2, 141–2
  and mechanical equivalent of heat 123
  and point defects 143–6, 167
  and recovery 90, 123, 142–6, 148, 150, 160–1, 164–5, 167
  and recrystallization 90, 123, 130, 143, 162–4, 165, 167
  and saturation level 50, 51, 169
  and stacking fault energy 154–5, 167–8
  and subgrain size 125–6, 140–1
  and substructure 132–8, 169
  in single crystals 49, 69–73, 151–3
  dissipation during deformation 165–6
  interpretation of 130–59
    historical survey 123–7
  measurement of 7, 16–22, 151, 168

SUBJECT INDEX

anisothermal annealing methods 19–21
isothermal annealing methods 21–2
reaction methods 22
single-step methods 16–9, 44
two-step methods 19–22,
mechanisms of 169
nature of 5
relation to
  density 117–8, 118–22, 138–40
  hardness 113–4, 118–22
  heat of formation (of intermetallic compound) 158
  resistivity 115–6, 118–22, 138–40, 143
release of 89–112
  immediately after deformation 18, 19, 61, 110–2, 154, 155, 162
  see also Release of the stored energy
spread of values 23, 44–7
summary of measured values of 24–45, 112, 115
theories of 122, 123–7, 130–65
  based on
    functional relations 124–5
    lattice imperfections 126–7, 130–65
    lattice strains 125–6
    phase change 123–4
Stored energy/expended energy ratio 6, 23, 49–51, 54–6, 69–74, 77, 81–2, 86, 124, 132, 137–8, 145, 155, 166
see also Rate of energy storage
Strain 7–8
Strain energy 15, 126, 141–2
  of dislocations 131, 133–5, 143
Strain rate see Rate of deformation
Stress 7
Stress, flow see Flow stress
Stress-strain relationship 8
Structural features and stored energy 169
Structure of cold-worked metals 127–30, 169
Subgrains 140–2, 163
  see also Cell structure, Substructure and stored energy
Substructure and stored energy 132–8, 140–1
  see also Cell structure, Dislocation structure of cold-worked metals
Surface/volume ratio and stored energy 58–9, 61–2
Swaging, deformation by 8

Taylor lattice 126, 131, 134
Temperature of deformation 51–6, 93–6, 128, 143–6
see also under specific materials
Temperature rise, hypothetical, in relation to stored energy 125
Tension, deformation by 8, 9, 59, 61, 128
  in aluminum 56, 71–3,
  in copper 10, 23, 56, 62, 71–2, 74, 91, 96–7, 103, 113, 117, 149
  in copper-gold alloys 87
  in copper-nickel alloys 63
  in copper-zinc alloys 107, 116, 118
  in gold-silver alloys 69
  in iron 23

in nickel 10
in silver 72–3, 103, 132
in single crystals 18, 56, 69, 71–3
in steel 10, 23
Thermal conductivity 18, 146
Thermal diffusivity 18
Thermal effects of deformation 9–10
Thermodynamic aspects of deformation processes 6–16
Thermodynamic function change with deformation 10–2
Thermodynamics
  First Law of 6, 14, 16
  irreversible, and plastic deformation 14–6
  Second Law of 14
Thermoelastic effect 9, 10, 17
Tilt boundaries 129
Tin, stored energy in 77, 79
Torsion, cyclic see Cyclic deformation
Torsion, deformation by 8, 18, 59, 61, 63, 74, 75, 96, 146
  in aluminum 105
  in arsenical copper 51
  in bismuth-thallium alloy 88
  in copper 23, 50–1, 54, 58, 81, 100, 149
  in copper-gold alloy 52
  in copper-zinc alloys 80–1, 107, 114, 116, 157
  in gold-silver alloys 52, 57, 109, 116, 120
  in nickel 51, 92, 115, 117, 121
  in silver 103
  in silver-magnesium alloys 87–8
Twins and twin faults 12, 128, 147–8, 169
  contribution to stored energy 167–8
Twist boundaries 129, 142

Vacancies 13, 14, 131, 144, 148–9, 154, 163
  see also Point defects
Vanadium, dislocation structure in 129
Volume change on deformation 7
  see also Density

Wire drawing, deformation by 8, 12, 18, 23, 59–61, 128
  in aluminum 117
  in copper 54, 117
  in copper-gold alloys 85–6
  in gold 122
  in gold-silver alloys 47, 52, 56–7, 62–3, 66, 113, 164
  in iron 132, 141
  in silver-magnesium alloys 156
Work
  friction 8, 16, 61
  of deformation 7–8, 16–7, 18, 66
  redundant 8, 61, 146
Work hardening 6
  and rate of energy storage 138
Work integral, evaluation of 8
Work softening 7, 62–3, 113, 129, 146, 149

X-ray data   125–8, 140–2, 146–7, 156, 158
   dislocation densities from   142, 147

Zinc
   release of energy from   106, 160

Zirconium
   release of energy immediately after deformation   110–1, 162
   stored energy in   77–80
      relation with flow stress   119
   stored energy/expended energy ratio   51

CONTENTS OF PREVIOUS VOLUMES IN THE SERIES

VOLUME 1

Progress in the Theory of Alloys   G. V. RAYNOR
Theory of Dislocations   A. H. COTTRELL
Crystal Boundaries   R. KING and B. CHALMERS
Age Hardening of Metals   G. C. SMITH
Hardening Response of Steels   E. H. BUCKNALL and W. STEVEN
Preferred Orientation in Non-Ferrous Metals   T. LI. RICHARDS
Diffusion of Metals in Metals   A. D. LE CLAIRE

VOLUME 2

Order–Disorder Changes in Alloys   H. LIPSON
Rate Processes in Physical Metallurgy   I. I. BETCHERMAN
Anisotropy in Metals   W. BOAS and J. K. MACKENZIE
Developments in Magnesium Alloys   H. G. WARRINGTON
Internal Strains and Recrystallization   R. W. CAHN
Researches on the Polygonization of Metals   A. GUINIER and J. TENNEVIN
Polygonization in Strongly Deformed Metals   C. CRUSSARD, F. AUBERTIN, B. JAOUL and G. WYON

VOLUME 3

Crystallography of Transformations   J. S. BOWLES and C. S. BARRETT
Properties of Metals at Low Temperatures   D. K. C. MACDONALD
Recent Advances in the Electron Theory of Metals   N. F. MOTT
Twinning   R. CLARKE and G. B. CRAIG
Ferromagnetism   URSULA M. MARTIUS
Quantitative X-ray Diffraction Observations on Strained Metal Aggregates   G. B. GREENOUGH
Recrystallization and Grain Growth   J. E. BURKE and D. TURNBULL
Structure of Crystal Boundaries   B. CHALMERS

VOLUME 4

Internal Friction in Metals, A. S. NOWICK
The Mechanism of Oxidation of Metals and Alloys at High Temperatures   KARL HAUFFE
Gases in Metals   C. R. CUPP
The Theory of Sintering   G. A. GEACH
Theory of Dislocations   A. H. COTTRELL
Diffusion in Metals   A. D. LE Claire
Nucleation   J. H. HOLLOMON and D. TURNBULL

VOLUME 5

The Fracture of Metals   N. J. PETCH
Geometrical Aspects of the Plastic Deformation of Metal Single Crystals
    R. MADDEN and N. K. CHEN
The Structure of Liquid Metals   B. R. T. FROST
Report on Precipitation   H. K. HARDY and T. J. HEAL
Solidification of Metals   URSULA M. MARTIUS

VOLUME 6

The Effect of Hydrostatic Pressure on the Electrical Resistivity of Metals   A. W. LAWSON
The Filamentary Growth of Metals   H. K. HARDY
The Austenite: Pearlite Reaction   R. F. MEHL and W. C. HAGEL
Recent Advances in Knowledge Concerning the Process of Creep in Metals   A. H. SULLY
The Mechanism of Evaporation   O. KNACKE and I. N. STRANSKI
Mosaic Structure   P. B. HIRSCH

VOLUME 7

Equilibrium Diffusion and Imperfections in Semiconductors   J. N. HOBSTETTER
The Physical Metallurgy of Titanium Alloys   R. I. JAFFEE
Thermodynamics and Kinetics of Martensitic Transformations   LARRY KAUFMAN and MORRIS COHEN
The Stored Energy of Cold Work   A. L. TITCHENER and M. B. BEVER
The Properties of Metals at Low Temperatures   H. M. ROSENBERG

VOLUME 8

Work Hardening of Metals   L. M. CLAREBROUGH and M. E. HARGREAVES
Grain Boundaries in Metals   F. WEINBERG
X-ray Studies of Deformed Metals   B. E. WARREN
Substructures in Crystals Grown from the Melt   C. ELBAUM
Defects in Pure Metals   W. M. LOMER

VOLUME 9

Nuclear Magnetic Resonance in Metals   T. J. ROWLAND
The Effect of Temperature and Alloying Additions on the Deformation of Metal Crystals
   R. W. K. HONEYCOMBE
Effects of Environment on Mechanical Properties of Metals   IRVIN R. KRAMER and LOUIS J. DEMER
The Hydrogen Embrittlement of Metals   P. COTTERILL
The Structure and Properties of Solid Solutions   J. M. SIVERTSEN and M. E. NICHOLSON

VOLUME 10

Alloy Phases of the Noble Metals   T. B. MASSALSKI and H. W. KING
Mechanisms of Growth of Metal Single Crystals from the Melt   D. T. J. HURLE
Precipitation Hardening   A. KELLY and R. B. NICHOLSON
Surface Diffusion   J. M. BLAKELY

VOLUME 11

Condensation and Evaporation: Nucleation and Growth Kinetics   J. P. HIRTH and G. M. POUND

VOLUME 12

The Fracture of Metals   JOHN R. LOW, JR.
Eutectic Alloy Solidification   G. A. CHADWICK
The Effect of Metallurgical Variables on Superconducting Properties
   J. D. LIVINGSTON and H. W. SCHADLER

VOLUME 13

The Mechanical Properties of Ordered Alloys   N. S. STOLOFF and R. G. DAVIES
Binding of Solute Atoms to Dislocations   NICHOLAS F. FIORE and CHARLES L. BAUER
Fracture Toughness: An Examination of the Concept in Predicting the Failure of Materials   P. KENNY
   and J. D. CAMPBELL
Metallurgy of Meteorites   H. J. AXON
The Engel–Brewer Theories of Metals and Alloys   W. HUME-ROTHERY
Crystal Structure Determination by Electron Diffraction   J. M. COWLEY
Mechanical Behaviour of Crystalline Solids at Elevated Temperature   OLEG D. SHERBY and PETER M. BURKE

VOLUME 14

Recent Progress in Metallurgical Thermochemistry   O. KUBASCHEWSKI and W. SLOUGH
The Stability of Metallic Phases   L. KAUFMAN
The Electronic Structure of Pure Metals
   Part A. Electron Theory of Pure Metals   W. M. LOMER
   Part B. Fermi Surfaces and Physical Properties of Some Real Metals   W. M. LOMER and W. E. GARDNER
Liquid Metals and Vapours under Pressure   R. G. ROSS and D. A. GREENWOOD

VOLUME 15

Part 1. The Growth and Structure of Eutectics with Silicon and Germanium   A. HELLAWELL
Part 2. Topologically Close-packed Structures of Transition Metal Alloys   A. K. SINHA

VOLUME 16

High-Angle Grain Boundaries   H. GLEITER and B. CHALMERS

~~Periodicals~~
v.17
1973

TN
1
P76
v.17